Brad Bergan

SPACE RACE 2.0

SpaceX, Blue Origin, Virgin Galactic, NASA, and the Privatization of the Final Frontier

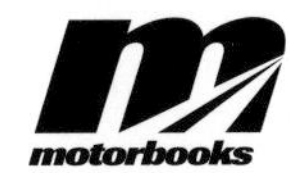

CONTENTS

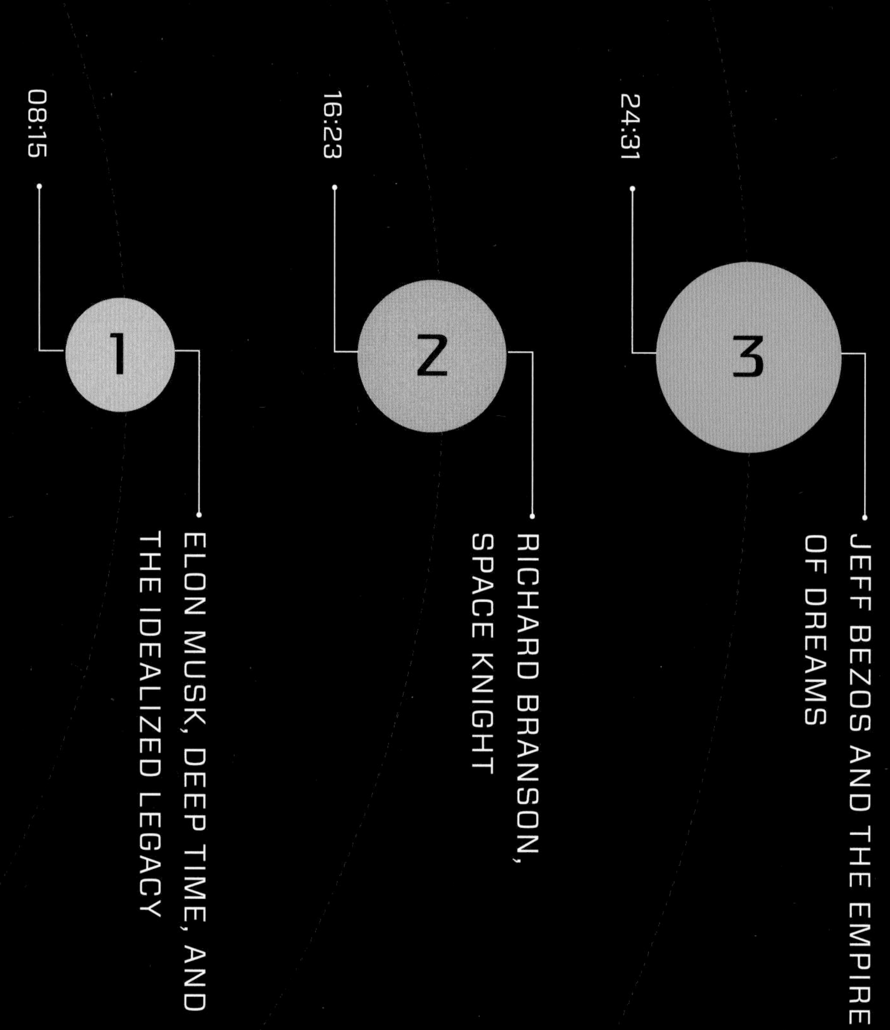

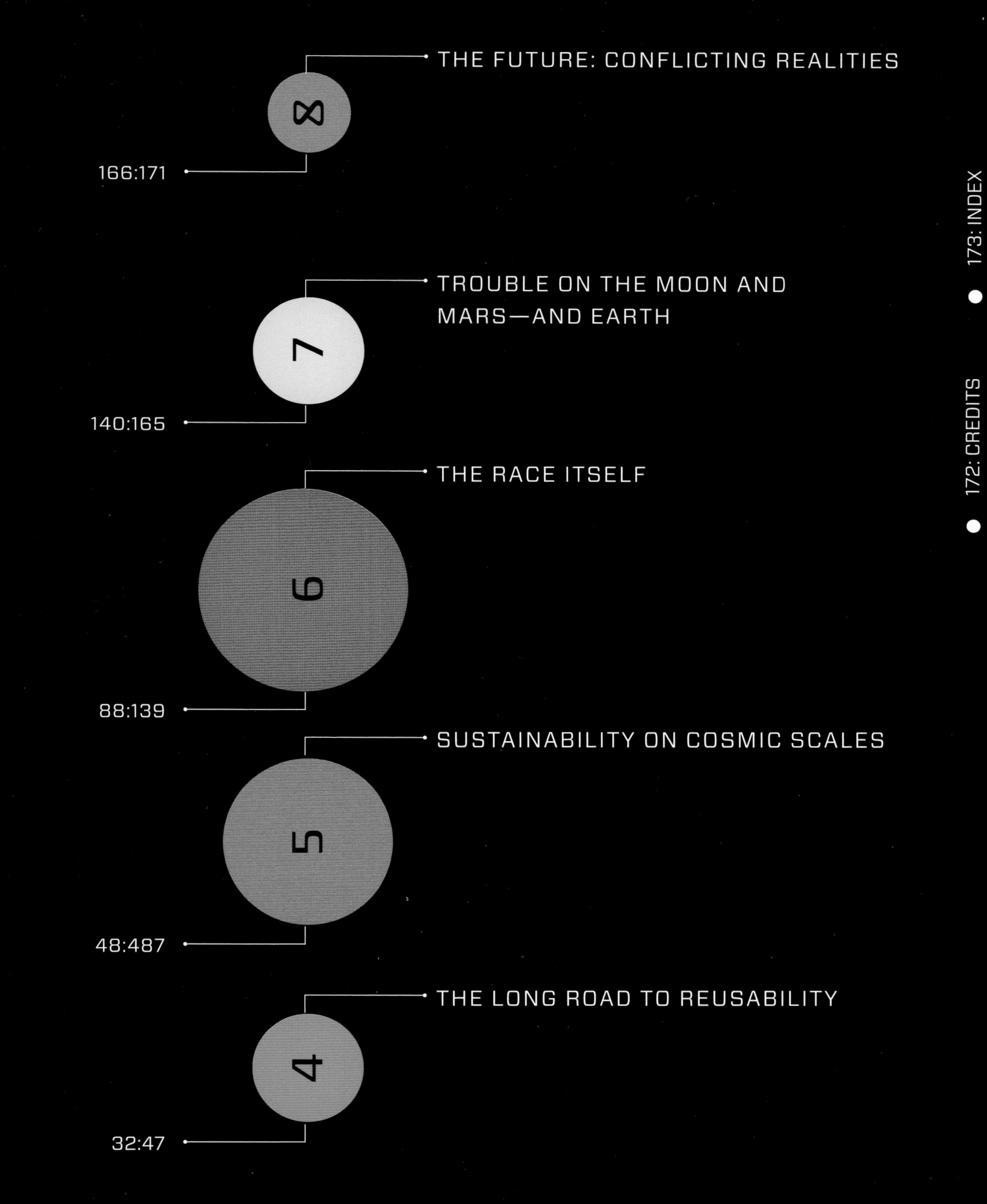

BLUE ORIGIN

Dawn broke over the desolate flatlands before the mountainous horizon of West Texas as the richest man in the world prepared to launch himself into space under a pale blue sky. Within Blue Origin's Van Horn facility, the countdown was eventually drowned out as the New Shepard's Blue Engine 3 rockets flared to life, mixing liquid hydrogen and liquid oxygen into a downward explosion of unconscionable force, lifting the founder of Amazon and Blue Origin, CEO Jeff Bezos; his brother Mark Bezos; an eighty-two-year-old aviator; and an eighteen-year-old boy to the edge of space.

Social media was already abuzz with excitement for the launch, but this was soon matched with snarky commentary about the oddly phallic-shaped spacecraft. And yet, that strange-looking rocket was also emblematic of the radical shift in the evolution of crewed space exploration. For decades, the conventional visions involved idealistic promises of sci-fi futures like those portrayed in iconic series like *Star Trek*. But the reality of space travel in the twenty-first century is the result of a strategic arrangement of industries with supply chains encircling the planet (like Bezos's immense corporate superpower), the prowess of novel advanced computing and AI-assisted software, and unprecedented public-private partnerships that allow billionaires like Elon Musk, Richard Branson, and Bezos to subsidize their dreams of extraterrestrial trade and market domination.

As one of a handful of billionaires with sights on settling the moon for commerce and exploration, Bezos's flight was, in a sense, propelled by the words of another billionaire: Elon Musk. "One path is we stay on Earth forever, and then there will be some eventual extinction event," the SpaceX CEO said at a 2016 conference in Mexico.

"I do not have an immediate doomsday prophecy, but eventually, history suggests, there will be some doomsday event. The alternative is to become a spacefaring civilization and a multi-planet species, which I hope you would agree is the right way to go."

But while survival seems to motivate Musk's drive to revive human space exploration and manifest its multi-planet destiny, the drive to space among Musk, Bezos, and Branson resembles not the familiar dreams of sci-fi cinema, nor the twentieth-century Space Race prompted by the Cold War and the military-industrial complex. Instead, Space Race 2.0 has more in common with another indispensable phase of American history, from a time before radio.

In the latter part of the nineteenth century, the consensus was the world was changing, perhaps too fast to adapt. And society might not get a second chance to build relations "between the palace of the millionaire and the cottage of the laborer," before a "relapse to old conditions . . . would sweep away civilization," wrote Andrew Carnegie in his 1889 article for the *North American Review* titled "The Gospel of Wealth."

In it, he argued for the primacy of philanthropy in wealth-building among the richest of his age in order to continue to lift the standard of living so that the progress of the past century would not be lost. "The poor enjoy what the rich could not before afford," wrote Carnegie. "What were the luxuries have become the necessaries of life. The laborer has now more comforts than the landlord had a few generations ago." This defense of industrial expansion as a force for rising standards of living forged the way for nascent technology, production, and wealth-building, juxtaposing them to sentiments of noble sacrifice and goodwill on the parts of the world's most privileged and powerful for everyone else. In this light, Elon Musk's drive to transform humanity into a two-world species, along with Jeff Bezos's "epiphany" that we should save Earth from climate change and pollute space instead, are not so different from Carnegie's philanthropy.

In a way, we never really left the nineteenth century. Back then, a group of Gilded Age industrialists leveraged their wealth by investing in technology, materials, and infrastructure to become captains of their respective industries. John D. Rockefeller founded Standard Oil Company, which came to control 90 percent of pipelines and refineries in the United States by the 1880s. Like Carnegie, he emphasized philanthropy, giving more than $500 million in donations throughout his life and work. John Pierpont Morgan used his family's wealth to dominate half of the railroads in the country. At the same time, he used his vast resources to invest in promising minds like Thomas Edison and his electricity company. None of them overmatching the others, they seeded the future, providing new avenues of employment in jobs with unforeseen levels of productivity and efficiency, despite their relative apathy for workers.

The second Space Race is inherently tied to a budding technological revolution of the 2020s, but, speaking financially, the success of the nineteenth-century "robber barons" has yet to be outdone to this day. Adjusted for inflation, Rockefeller's net worth [circa 2021] was more than $400 billion. By early 2022, after a global pandemic that saw substantial growth for the wealth of billionaires, Elon Musk's net worth was roughly $236 billion.

In form, Elon Musk, Jeff Bezos, Bill Gates, Richard Branson, and many other billionaires are just like those Gilded Age robber barons. The difference lies only in the message and materials. Where Gilded Age industrial leaders manufactured steel, drilled for oil, and constructed the railroad infrastructure necessary to modernize the United States, the robber barons of Space Race 2.0 have stripped down rocket designs to seek a more efficient means to bring the world economy into space, to give birth to space tourism, and to settle the moon and Mars not for one nation or race but for twenty-first-century society—from its highest-level science and noblest aims to the messy realities of wealth inequality and harsh working conditions. The grueling demands of working in or around a nascent space industry are unspeakably challenging, especially when compounded by the onslaught of climate change and a global pandemic. But the pace of Space Race 2.0 is and will be relentless, as countless scientists, engineers, and politicians join hands with a few billionaire space barons to signify humanity's first solid steps into a wider universe.

3...

2...

1...

1

ELON MUSK, DEEP TIME, AND THE IDEALIZED LEGACY

Elon Musk was born to Maye and Errol Musk in South Africa's Pretoria on June 28, 1971. He lived there for approximately the first eight years of his life, playing with explosives, building rockets, and reading incessantly. "I was raised by books," said Musk in a 2017 *Rolling Stone* interview.

Today, we know Musk as a tech billionaire at the helm of several heavily subsidized firms developing, designing, and producing novel vehicles, including all-electric vehicles for this world, and reusable rockets designed to take us to the next (the moon, Mars, and beyond). In a humble twist of casual equivocation, you could say Musk is a fan of world-building—the kind most prominent in a genre of fiction called "hard sci-fi": a classic default mode of the genre that emphasizes technology, space travel, and conventional philosophy in the unfamiliar context of theoretical physics. Some varieties of hard sci-fi employ maximally abstract, almost godlike macro perspectives on society. Musk likes this perspective, but it didn't come *ex-nihilo;* his favorite books growing up were Isaac Asimov's *Foundation* series, which centers on the revolutionary

Musk, at a public event in Germany, announcing plans to build his Model 3 and Model Y series EVs there

work of a mathematics professor in the distant future named Hari Seldon, who postulates a scientific means of forecasting the "deep future" of entire civilizations based on the behavior of crowds. In the novels, Seldon predicts a 30,000-year-long Dark Age for humanity, so he organizes the founding of multiple colonies on distant planets to keep the light of consciousness alive. Musk regularly echoes this sentiment today in defense-tweets of what he probably considers his raison d'être. Musk says the purpose of his budding tech empire is to build a launch system capable of lifting humans to the moon and Mars. Why? "To extend the light of consciousness," tweeted the billionaire in 2021.

"The lesson I drew from [reading Asimov and Gibbon's *Decline and Fall of the Roman Empire*] is you should try to take the set of actions that are likely to prolong civilization, minimize the probability of a dark age, and reduce the length of a dark age if there is one," surmised Musk in the *Rolling Stone* interview. Following his parents' divorce, Musk entered a personal dark age, where his father played prime mover. On the subject of Errol Musk, author of *Elon Musk: Tesla, SpaceX, and the Quest for a Fantastic Future* Ashlee Vance said his entire family shared a firm consensus: "They're in agreement that [Errol] is not

↑ The science fiction author and biochemist Isaac Asimov, one of the greatest science fiction authors and one of Musk's favorites

← Elon Musk immersed himself in "hard sci-fi" where often we're met with highly advanced technological civilizations. Painting by Anton Brzezinski

↑ | Errol Musk, Elon's father

← | Maye Musk, Elon's mother, at a fashion week event

→ | These Shantytown shacks are located in Soweto, near South Africa's mining belt. Many miners live out the majority of their lives stuck in similar environments and actively protest this. The weight of the knowledge of this being the cost of space legacy would be unbearable.

a pleasant man to be around but have declined to elaborate," she wrote.

While reportedly a millionaire before his thirties, Musk's father "was physically, financially and emotionally manipulative and abusive," according to Maye in her 2019 memoir: *A Woman Makes a Plan: Advice for a Lifetime of Adventure, Beauty, and Success.*

However, it's difficult to make the argument that a harrowing home environment spurred young Elon Musk to rise above his origins to achieve some measure of greatness, especially with not wholly unfounded claims that he benefited from financial privilege.

More specifically, the criticism alleges that Musk's success was largely due to a sizable inherited wealth: the riches from his father's emerald mine in apartheid South Africa. If this were straightforwardly the case, then it would mean that egregiously unethical working environments of countless Black mine workers are the reason Musk had enough to found and fund Tesla, SpaceX, and other companies, which would cast a dark shadow over the long road to power. This would turn his entrepreneurial and engineering success into a sobering example of the depths of cruelty that lie behind the successes many cherish as milquetoast advancements for humankind on Twitter feeds every day. But, this wouldn't be a major exception for space travel. After all, the United States and former Soviet Union's space programs were initially developed from the knowledge and rocket technology of German scientists, formerly of the defeated Third Reich.

But Musk the son directly addressed this talking point in 2019, vehemently denying a link between his success and alleged inheritance of Errol Musk's emerald money. "He didn't own an emerald mine & I worked my way through college, ending up ~100k in student debt," tweeted Musk. "I couldn't even afford a 2nd PC at Zip2, so programmed at night & [Zip2's] website only worked during [the] day." Musk's self-account is of a hardworking bootstrapper who forged his own way through precarious beginnings after separating from a dangerous father figure. But Musk's final retort protests too much: "Where is this bs coming from?" he added in his 2019 tweet. At the very least, it comes from logical necessity: the two events of Errol making a killing in a lucrative mining

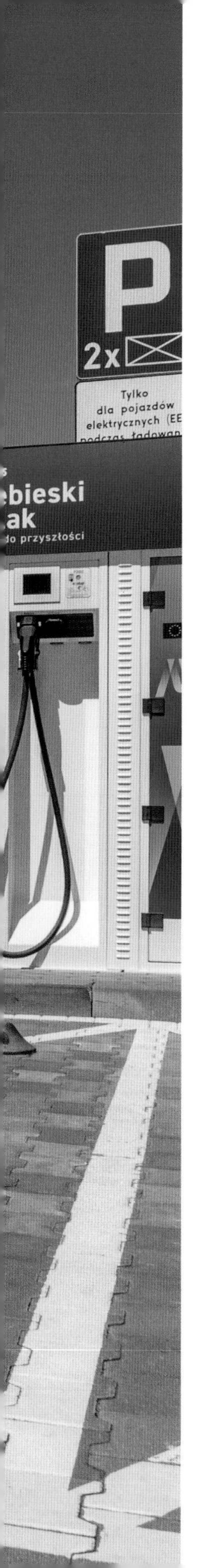

operation and his son Elon making a billionaire icon of himself "by the bootstraps" are not mutually exclusive. Not necessarily.

The Musk family "owned one of the biggest houses in Pretoria thanks to the success of Errol's business," which involved "large projects such as office buildings, retail complexes, residential subdivisions, and an Air Force base," wrote Vance in his book on Musk. However, even discounting the role of money in Elon Musk's formative years, some key advantages instilled an instinct for an engineering perspective on the world. For one, it turns out his dad was exceedingly intelligent. "[B]rilliant at engineering, brilliant," said Musk in the *Rolling Stone* interview. After his parents split, Musk moved in with his dad in a suburb of Johannesburg called Lone Hill. "I'm naturally good at engineering [and] that's because I inherited it from my father."

"What's very difficult for others is easy for me," continued Musk. "For a while, I thought things were so obvious that everyone must know this." To him, his "natural" engineering sense was inherited from Errol. Knowing how electric wiring powers a house or how a circuit breaker or direct or alternating current works was like knowing your way home. "[W]hat amps and volts were, how to mix a fuel and oxidizers to create an explosive. I thought everyone knew this." It's common to take one's own skills for granted, even and especially when they're rare. It's the mark of someone who has worked to improve themselves, rather than merely to beat someone else for fame, riches, or power, since the latter can impose an external limit on personal growth.

However, parting ways with Errol may have had a greater influence on Musk than is commonly known. "You have no idea about how bad," said Musk about Errol in the *Rolling Stone* interview. "Almost every crime you can possibly think of, he has done. Almost every evil thing you could possibly think of, he has done. Um . . ."

Eventually, after cutting ties with his father, Musk cashed a check for $22 million from selling his company, Zip2. With this money, he funded another firm, X.com, later known as PayPal. Most of us weren't familiar with PayPal until the late 2000s, but by then, Musk had already sold it to eBay for approximately $180 million. With an even bigger pile of money, he started three companies that would seed several core technologies of the fourth industrial revolution: SpaceX, the private aerospace company that started with $100 million; Tesla, Musk's all-electric automaker founded with an initial $70 million injection; and another $10 million business called SolarCity. He pocketed the rest of his PayPal profits a full ten years before SpaceX's first-ever reusable flight system made its first jump into history.

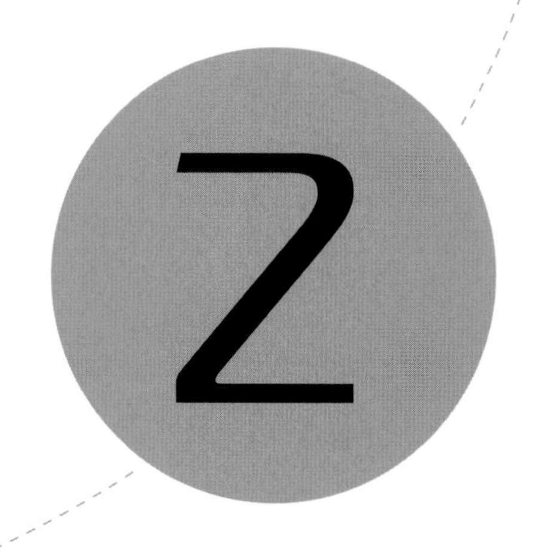

RICHARD BRANSON, SPACE KNIGHT

Branson crossed both the Atlantic and Pacific oceans in hot-air balloons. He's seen here (center) with Per Linstrand (left) and Steve Fossett (right) before taking off on their failed 1998 attempt to circumnavigate the globe in a balloon.

Richard Branson was born on July 18, 1950, in Surrey, London. He struggled with dyslexia as a child and didn't do well in school. But where he lacked a talent for numbers, he made great strides with people, discovering an inherent love of social life. While cutting deals at his first business venture, a magazine called *Student*, he made transactions via the name "Virgin" because he was "entirely new to business," according to Virgin's official website. Branson later opened a record shop in London, which helped him build enough capital to launch his own record label in 1972 that became a household name: Virgin Records. This legacy label recorded instant-hit operations like the Sex Pistols, Faust, and even the Krautrock band Can, all of whom found exposure to the global music industry through Branson's company. He sold his Virgin Records label to EMI in 1992 for a cool $683 million (£500 million).

In 1984, Branson founded Virgin Atlantic Airways, and in 1999, he founded Virgin Mobile, followed by Virgin Blue in 2000. He even invested in an expansion of the railway industry, and

"The moon landing was a cataclysmic moment for me."

—Richard Branson

Virgin Trains gained franchises for cross-country regions of British Rail, adding InterCity West Coast to his railway company. If you're beginning to notice a pattern here, you're not wrong. Branson's budding Virgin empire spread across such a diverse smattering of major industries you'd think his portfolio was hedging bets on capitalism. He started a Nigerian national airline and called it Virgin Nigeria and then tried to do the same in the United States in 2007. He even created a Virgin brand of soda and vodka, although they didn't really catch on. By 2021, Branson's Virgin Group included more than 400 companies that reached beyond thirty countries. But it was in the late 1980s that his daredevil business philosophy took off in the physical world of flight and engineering.

During the Reagan days of 1986, Branson crossed the Atlantic in his Virgin Atlantic Challenger II and then crossed both the Atlantic and Pacific via a hot-air balloon in 1987 and 1991, respectively. The British Royal family even knighted him in 1999, citing his exceptional service to the business world. In 2004, Sir Richard Branson founded Virgin Galactic, just two years after Elon Musk's SpaceX came into being.

In an interview released by Virgin Galactic, Branson explained that he,

↑ Richard Branson is the man behind Virgin Records, which brought the Sex Pistols to prominence.

← SpaceShipOne glides down for its approach to an airport in the Mojave Desert.

too, was moved by the 1969 moon landing of NASA's Apollo space program. "The moon landing was a cataclysmic moment for me. "But, while this instilled a profound desire to go there himself, he understood that it would take decades for technology to advance far enough for non-astronauts to venture into space. So, he "waited and waited" for the matrix of space-related technology to catch up to the dream of the first Space Race. Sadly, Congress reduced funding to NASA after serious tragedies like the explosive end of the Space Shuttle Challenger's 1986 liftoff, which killed all seven crew members.

Eventually, it became clear: If Branson was going to space, he'd have to find his own way there. So, he planned to use a variant of SpaceShipOne, a reusable suborbital spaceplane constructed by another company, Scaled Composites. This launch system was chosen after Virgin Group employee Alex Tai visited a Scaled Composites hangar and saw the space vehicle.

Originally conceived for private leisure spaceflights, it didn't take Branson long to decide this was the system he wanted to take him to space. SpaceShipOne was designed to be carried skyward via a carrier aircraft called White Knight, from which it would detach and rocket away toward the boundary of space. Using this design, Virgin Galactic planned to fly 3,000 passengers on SpaceShipTwo in just five years, each passenger paying $208,000. Within a decade, Virgin Group aimed to lift 50,000 passengers.

In July 2005, Scaled Composites and Virgin Galactic signed a deal to create The Spaceship Company, which would enable the construction of SpaceShipTwo and WhiteKnightTwo. By December of that year, New Mexico governor Bill Richardson revealed plans to spend $225 million on a spaceport and headquarters for Virgin Galactic near Truth or Consequences, New Mexico. The plans called for the conversion of twenty-seven square miles (7 m^2) of state land, and completion was slated for 2009 or 2010. In 2007, Virgin Galactic signed an agreement to lease 83,400 square feet (7748 m^2) of hangar space, in addition to terminals at Spaceport America, where the New Mexico Spaceport Authority (NMSA) would build and retain ownership of Virgin Galactic's facilities until it paid off a sum of $27.5 million over twenty years.

However, things took a tragic turn when a fiery explosion killed Charles "Glenn" May, Todd Ivens, and Eric Blackwell on July 26, 2007, all of

← SpaceShipTwo, the VSS Unity, debuted in Mojave, California, in 2014.

→ The VSS Enterprise, carried by the Virgin WhiteKnightTwo Mothership, performing a flyover

↓ The VSS Enterprise in flight

↑ Agents from the National Transportation Safety Board examine the wreckage of the VSS Enterprise.

→ A piece of debris from the VSS Enterprise

them Scaled Composites employees. The incident also landed three more employees in the hospital with life-threatening injuries. Virgin Galactic suspended all work on SpaceShipTwo for nearly a year to identify the precise cause of the accident. Eventually, engineers made changes to the nitrous oxide tank, and on January 23, 2008, the company finally debuted its SpaceShipTwo and WhiteKnightTwo designs. That summer, one year and two days after the fatal explosion at Mojave, the double-fuselage aircraft was renamed the Virgin Mother Ship (VMS Eve), ostensibly in homage to Branson's mother (although there are at least two other baffling readings of that name). It took flight for the first time on December 21, 2008.

After nearly two years of work, the VSS Enterprise finally took its first glide-flight on October 10, 2010, and by October 2012, Virgin Galactic had officially taken total ownership of The Spaceship Company, acquiring the remaining 30 percent stake that Scaled Composites had in the joint venture. On April 29, 2013, the VSS Enterprise made its first-ever powered test flight, executing a sixteen-second engine burn. At the same time, Branson declared that ticket prices for prospective flights would rise to $250,000, leading some to ask the reasonable question of who, specifically, would have deep enough wallets to enjoy the coming space tourism industry. Undaunted, Virgin Galactic's vehicle underwent two more powered test flights, one later in 2013

and a third in early 2014, but for twenty seconds each. The third test flight was meant to execute a longer, one-minute burn, but this was shortened due to worries about engine vibrations and oscillations.

And then tragedy struck once more. On October 31, 2014, when the VSS Enterprise was undergoing a rocket-powered test flight, it suddenly broke into pieces in the Mojave skies, killing the copilot, Michael Alsbury, and inflicting serious injuries to the pilot, Peter Siebold, who escaped after deploying his parachute. Later, the National Transportation Safety Board (NTSB) investigating the incident found that Alsbury had mistakenly unlocked the "feathering" reentry system far too early in the test flight.

The "feather" is supposed to deploy during reentry into Earth's atmosphere to provide a relatively smooth descent. Alsbury was supposed to unlock the feather in the middle of the powered ascent to ensure that the locks would release; if not, the vehicle would have to abort the entire flight and return to Mojave Air and Space Port. But, with the feather unlocked prematurely, there was little time to properly abort as a rapid shift in the distribution of aerodynamic forces caused the vehicle's twin tail booms to deploy into a configuration not capable of withstanding the air resistance of rapid ascent, ending in catastrophe.

But amid Virgin Galactic's trials and its first $4.7 million NASA contract to lift more than a dozen satellites into orbit, another major contender in the second Space Race was coming into maturity—one founded four years before Virgin Galactic and headed by the billionaire rival to Elon Musk: Jeff Bezos.

BLUE ORIGIN

JEFF BEZOS AND THE EMPIRE OF DREAMS

Renowned entrepreneur and e-commerce pioneer Jeff Bezos was born in Albuquerque, New Mexico, on January 12, 1964. His then-teenage parents, Jacklyn Gise Jorgensen and Ted Jorgensen, divorced after a year, and his mother married Mike Bezos, an immigrant from Cuba.

As a child, Bezos was fascinated by the mechanics of the world around him, especially during many cherished summers at his grandfather's Texas ranch. "His grandfather sparked and indulged Jeff's fascination with educational games and toys, assisting him with the Heathkits and other paraphernalia he constantly hauled home to the family garage," according to a profile of Bezos in a 1999 *Wired* report.

Heathkits were DIY kits that kids could use to build electronic products, and Bezos took this knowledge home with him, converting his parents' garage into a personal laboratory and experimenting with the electrical circuits around the house.

"Picture the scattered components of a robot; an open umbrella spine clad in aluminum foil for a solar cooking experiment; an ancient Hoover vacuum cleaner being transformed into a primitive

Jeff Bezos (second from right) with his July 2021 Blue Origin crew: Oliver Daemen (left), Wally Funk (second from left), and brother Mark Bezos (right)

hovercraft," continued the *Wired* profile. Most kids who dream about becoming an astronaut grow out of it by their teens, electing instead to read or watch movies about space or sci-fi. And if Bezos had similar dreams, they were redirected toward an entrepreneurial parallel: Instead of the conventional military-ace-to-NASA pipeline, he'd just make a mountain of money and do it himself. By his teens, Bezos was in Miami, where he graduated valedictorian of his high school, and this is when he started his first company: the Dream Institute, an educational summer camp for fourth-, fifth-, and sixth-grade students. Later, he studied computer science and electrical engineering at Princeton University, where he graduated summa cum laude in 1986. With Ivy League credentials in hand, he went to work on Wall Street.

Several companies employed Bezos on Wall Street, including FITEL, Bankers Trust, and D. E. Shaw, where, by 1990, he had soared through the ranks to become the youngest senior vice president at the investment firm. But Bezos had larger ambitions, which is why he quit in 1994. Given the considerable financial comfort afforded by his position, this was unusual. Even risky. And he did it to wade into the budding world of e-commerce, moving to Seattle to start what began as an online bookstore, but in the next decade grew to become one of the most powerful economic forces in the world: Amazon.com. And this was just the beginning of the Bezos empire.

Bezos founded Amazon on July 16, 1995, starting with a few employee programmers designing the website in his garage, and later expanding into a two-bedroom house. Despite these modest beginnings, the company began to grow at breakneck speeds. With zero press promotion, it sold books throughout the United States and forty-five other countries in its first thirty days, with sales blasting off to $20,000 per week after two months. Even Bezos was surprised. Of course, all of this wouldn't have meant much for space travel if Amazon hadn't survived the dot-com bubble crash of the late 1990s. Amazon not only avoided going bust but grew by several orders of magnitude, leaping from $510,000 in revenue in 1995 to more than $17 billion by 2011.

Bezos even bought the *Washington Post* in 2013, acquiring one of the most powerful media companies in the country. In 2017, Amazon acquired Whole Foods, a popular urban grocery store that specializes in organic and vegan food. By 2021, Amazon formally declared that Bezos would step down as CEO in the third quarter of that year. All of this is to say that without even involving space travel, Bezos and his business empire were already woven into the economic fabric of the most powerful nation in the world. And, he had a secret project in the works.

Since founding Blue Origin in 2000, Bezos mostly kept his personal aerospace firm under wraps, giving only intermittent disclosures required by both NASA and the Federal Aviation Administration, while pursuing additional funding and necessary regulatory permissions. By early 2010, the public knew that Bezos was developing Blue Origin's New Shepard rocket, but facts were scarce: It would eventually lift three or more astronauts to suborbital altitudes, and the firm had received $2.7 million from NASA to develop an astronaut escape system, in addition to a space capsule prototype composed of composite materials for upcoming structural tests. "If we're famous for anything . . . it's for being quiet," said the company's lead engineer, Gary Lai (later senior director of program management), during a Next-Generation Suborbital Researchers Conference in Boulder, Colorado, in the early 2010s.

"One of the reasons is [that] it certainly keeps our marketing and public relations staff small," quipped Lai. This came roughly three years after several test launches of Blue Origin's Goddard rocket in November 2006, an early development vehicle for the New Shepard program, which would become a vertical takeoff and landing (VTOL) system capable of routinely lifting humans to space at competitive prices. In November 2009, the company selected three research payloads to hitch a ride on the suborbital vehicle: a Microgravity Experiment on Dust Environments in Astrophysics (MEDEA), managed by the University of Central Florida's Joshua Colwell; a 3D critical wetting experiment in microgravity led by Purdue University's Stephen Collicott; and Effective Interfacial Tension Induced Convection (EITIC), whose principal investigator was John Pojman of Louisiana State University.

The New Shepard was far larger than the Goddard vehicle, with much expected of it. Named after the first American in space, Alan Shepard, Blue Origin's first mission-capable rocket also harkened back to NASA's Project Mercury, which lifted Shepard to suborbital altitudes aboard the Freedom 7. But Bezos's firm differs from NASA's Mercury project, which employed an eighty-three-foot (25.5 m) single-stage rocket for suborbital lift. Called the Mercury-Redstone Launch Vehicle, NASA's first crewed Space Race vehicle burned a mixture of liquid oxygen and alcohol to generate about 75,000 pounds-force (334 kN) of thrust. Notably, this design was a direct descendent of the German V-2 rockets that pummeled London during the Blitz in World War II.

The New Shepard generates power with a BE-3 liquid hydrogen and liquid oxygen engine and stands fifty-nine feet (18 m), much shorter than the Mercury-Redstone. The New Shepard's crew capsule features large windows that offer passengers a breathtaking view of Earth at the apex of its ballistic trajectory. After separating from the single-stage rocket, the crew capsule continues toward its highest position while the booster returns to the surface via guided vertical landing. Now capable of holding six astronauts, the crew

← Aerial view of the Amazon Spheres at its Seattle headquarters and office towers

→ (Left) Pioneer rocket scientist Robert H. Goddard, for whom the Blue Origin rocket was named

→ (Right) Alan Shepard, the first U.S. citizen to reach space and the namesake of Blue Origin's first space-worthy rocket

Blue Origin's New Shepard rocket, launching

capsule also experiences a few moments of weightlessness before falling back into the atmosphere, using a hybrid parachute and retro-rockets system to make a controlled, soft touchdown.

And, after fifteen years of mostly secretive development, Blue Origin's New Shepard vehicle executed its first-ever successful launch and landing.

At 12:21 p.m. EST on November 23, 2015, the New Shepard rocket delivered 110,000 pounds (490 kN) of thrust, achieving a speed of Mach 3.72 (3.72 times the speed of sound) and reaching a max altitude of 329,839 feet (100.5 km). While the crew capsule detached and made its own landing without incident, the real challenge came when it was time to attempt a vertical landing of the booster. "Perfect landing," said mission control at Blue Origin's West Texas launch facility as the rocket landed gently with a leisurely speed of 4.4 mph (7.1 km/h), an event livestreamed on YouTube. "We made history today. Now, who wants to go to space?" Blue Origin had become one of the first entities on Earth to execute a successful vertical takeoff and landing of a rocket, a feat that many considered purely science fiction until then. Bezos posted his first-ever tweet, affirming the historic event, and making waves across the depths of the Twitterverse.

Of course, New Shepard wasn't the first rocket to carry out a successful vertical takeoff and landing. Back in 1996, the DC-X from McDonnell Douglas rose to roughly 10,200 feet (3,109 m) and then touched down without a snag. And SpaceX's Grasshopper rocket had ascended to 2,441 feet (744 m) in 2013, landing without incident in Central Texas. But the gravity of Bezos's accomplishment was short-lived, due to what some might call the opening act of space baron drama.

↑ The New Shepard booster, which performed a vertical landing after launching the crew capsule

→ Artist concept of the McDonnell Douglas DC-X (Delta Clipper-Experimental) Reusable Launch Vehicle (RLV), which performed a successful VTOL in 1994

NASA

THE LONG ROAD TO REUSABILITY

Despite beating Jeff Bezos to the VTOL punch, SpaceX CEO Elon Musk, whose Falcon 9 rockets were still in development by late November 2015, did what he does best online. He trolled. "Congrats to Jeff Bezos and the Blue Origin team for achieving VTOL on their booster," tweeted the billionaire on November 24. To develop space travel at scale, the emphasis had shifted from highly sophisticated NASA engineering to a more minimal approach: low-cost, reusable rockets. The very same day, Musk quickly followed up his salutary patronization with another tweet, disputing the significance of Blue Origin's successful VTOL. "It is, however, important to clear up the difference between 'space' and 'orbit', as described well by "What If", an online blog that breaks down common questions about scientific ideas. Musk argued that by flying only into suborbital space, Bezos had avoided the need to accelerate his rocket to the high velocities required to achieve orbit around Earth.

SpaceX's Falcon 9 makes its first successful upright landing at sea on April 8, 2016, some 200 miles (322 km) offshore after launching from Cape Canaveral.

SpaceX had attempted to land its Falcon 9 booster on a sea-based platform without success (the first successful landing of Musk's booster would happen December 2015, with a landing at sea on a drone ship in April 2016). But in his tweet, Musk implied that Blue Origin's New Shepard vehicle was sorely outclassed by his Falcon 9 rocket, which had already lifted payloads into orbit, some even into geostationary orbit, approximately 22,000 miles (35,406 km) above the surface of the planet. Musk then opened a Twitter thread, where he explained the contextual physics of rocketry and clearly stated that his Falcon 9's nine Merlin engines possessed far more power than the New Shepard's BE-3 engine. And when it came to sheer thrust, he was right: Just one Merlin engine generates roughly 210,000 pounds-force (934 kN) of thrust, whereas a single BE-3 in the New Shepard vehicle maxed out at 110,150 pounds-force (490 kN).

In other words, Musk was bragging about the size of his . . . rocket.

Bezos gave a three-lined retort during a media teleconference following Blue Origin's first successful VTOL. First, Musk's Falcon 9 booster doesn't achieve orbit, either. In this light, SpaceX, too, had yet to surpass the ballistic threshold. The Falcon 9 rocket also executes a deceleration burn in space to lower its reentry velocity, which means New Shepard endures a more challenging (read: tumultuous) reentry window. Last, Bezos argued, "The hardest part of landing is probably the final landing segment." The key to Bezos's comeback jab was the subtext, where you can hear him speaking at Musk's level. By "final landing segment," Bezos referred to SpaceX's issues with keeping its rockets erect after landing.

To be clear, Bezos was saying that Musk can't keep it up.

While this billionaire burn would become the first of many between Bezos and Musk, it runs deeper than phallic in-jokes. The development of truly reusable rocket systems was a long, hard struggle, bristling with failures and renewed ambition. It's easy to understand why Bezos kept Blue Origin's roller-coaster development flow behind closed doors, but if there's one everlasting feature of Musk's personality, it's the contrast between his soft-spoken, introverted demeanor and his persistent love of the spotlight. In line with his uncommon dualism, the story of SpaceX's rocket development was less hidden than quiet, less braggadocio than ironic self-affirmation. When Elon Musk filed a patent in 2004 (after its 2002 founding) for a crucial engine component that requires no replacement post-flight, it became clear that his contribution to the second Space Race would be to pilot its core concept: reusable rockets, which "are very desirable designs from a cost reduction standpoint," wrote his company in its U.S. patent.

Musk and his engineers worked for nearly a decade before the first test vehicle, the Grasshopper, was ready for live fire tests. At 102 feet (31 m) tall, the prototype was constructed around an empty tank that served as the progenitor of the Falcon 9 rocket that had already finished a qualification program five months earlier. Erring on the side of caution, Musk wanted to try a VTOL

← Jeff Bezos, left, announces the Blue Origin rocket in September 2015 as Florida governor Rick Scott applauds during a press conference at Cape Canaveral.

→ Falcon 9 lifts off from Cape Canaveral on April 18, 2014, carrying its Dragon CRS-3 spacecraft on a resupply mission to the International Space Station.

The Orbital Sciences Corporation Antares rocket, with the Cygnus spacecraft onboard, explodes moments after launch from the NASA Wallops Flight Facility on October 28, 2014.

with the smaller (and less financially risky) Grasshopper before the Falcon 9, and the former made its first hop into the future of space travel on September 21, 2012. It didn't ascend higher than half a mile (805 m), but this was a limit imposed by the Federal Aviation Administration, which granted the license for SpaceX's test program.

Test flights continued at a relatively small launch facility in McGregor, Texas, and enabled Musk and his company to collect precious data to enhance their rocket technology for future experiments. In the early 2010s, nothing about SpaceX's hardware was revolutionary since it was highly similar to McDonnell Douglas's rockets from the 1990s. But SpaceX wasn't skipping any steps in its journey to develop the software needed to guide a rocket not only through launch but through the countless calculations and highly precise timing of controlled descents to Earth. The daunting challenge of building an autonomous launch system demanded every step be tested by brute force, and the Grasshopper's last flight happened in October 2013.

Next up was the main course: the Falcon 9 reusable rocket (F9R), designed specifically to fly higher and hover in place. The prototype was constructed around a 131-foot (40 m) tank of the second-gen version of the vehicle and featured unique retractable landing legs that would become a staple of SpaceX landings. In May 2014, it flew 0.62 miles (1 km) into the air before landing on the launchpad. But this minor success was followed by a mishap that August. Following a nominal takeoff, the Falcon 9 veered off course. Since these test launches happen unconscionably close to rural or residential areas, the rockets are equipped with automated software that causes them to self-destruct if they leave the designated test area. And that's what happened with the August 2014 launch.

The August test flight's error involved a blocked sensor. One SpaceX official said a fully operational rocket would

↑ Here's an up-close view of the Falcon 9's retractable legs. The first orbital-class rocket to perform a successful return to launch site and vertical landing is on permanent display outside SpaceX headquarters in Hawthorne, California.

← Shown here are the retractable legs in action. Falcon 9 arrives at Landing Zone 1 in Cape Canaveral after delivering NROL-76 to orbit on May 1, 2017.

→ Falcon 9 rocket stands after making its first successful upright landing on the Of Course I Still Love You drone ship on April 8, 2016.

MARMAC 304

come with a redundant backup. While no one builds a space rocket with the hopes that it could become a cool fireworks display, a lot can go wrong. In the early days of the first Space Race between NASA and the USSR space program, astronauts and cosmonauts looked on with visceral dread as several prototypes of their eventual rides exploded on the launchpad shortly after takeoff or slammed back into the ground after flipping over. As recently as 2014, a NASA Antares launch experienced terminal failure when the rocket exploded before it could reach the sky. But one advantage SpaceX and its rivals had over the first Space Race's protagonists was the lessened urgency to catapult humans into space, which meant that tests could double as operational flights with cargo aboard.

"SpaceX was built on 'test, test, test, test, test,'" a SpaceX engineer told a NASA interviewer in 2013. "We test as we fly." This line is endemic to the core concept of rocket reusability: a steady flow of satellite and cargo launches to generate income, which in turn fuels experimental sojourns into nascent technology, like first-stage rocket recovery, in addition to testing other items of new equipment like grid fins and landing legs. For safety's sake, the rockets typically made their ascent over the open ocean cleared by the U.S. Coast Guard, minimizing the risk to populations living below. In the twenty-first century, there's a significant link between reusability and the view of sustainability in the context of a rapidly warming climate, but before we dive into the image and symbolism of SpaceX, its primary accomplishments demand attention.

The first official commercial mission of SpaceX happened on September 23, 2013, and involved the lofting of a

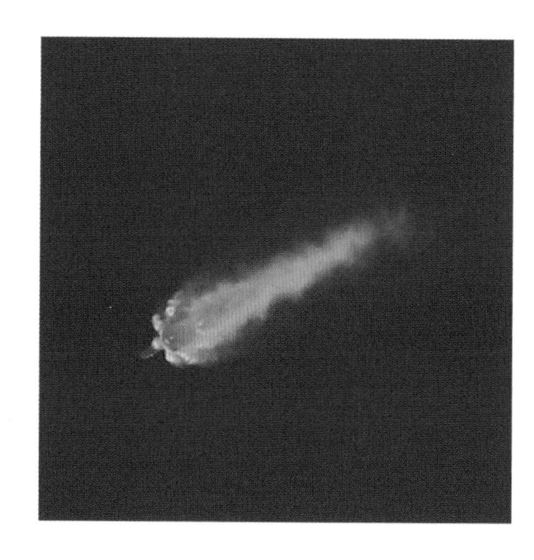

↑ A SpaceX Falcon 9 rocket on its seventh official Commercial Resupply Services (CRS) mission to the orbiting International Space Station breaks apart on Sunday, June 28, 2015, after launching from Launch Complex 40 at the Cape Canaveral.

↓ Falcon 9 first stage arrives in Port Canaveral, Florida, on April 4, 2017. Aboard its landing platform ship is the autonomous spaceport drone ship named Of Course I Still Love You.

Canadian satellite dubbed CASSIOPE. The satellite was safely delivered into orbit, and the firm's engineers tested the first-stage booster's landing capabilities over the ocean before it lost control and plunged into the deep blue waters. The next year, on April 18, 2014, and one day following the company's first-ever F9R test, SpaceX launched the CRS-3 atop a Falcon 9. This was the first operational F9R equipped with the retractable landing legs. On its descent to the ocean, the first-stage rocket executed an excellent controlled landing, but instead of a touchdown on a floating platform, the rocket splashed into the water. This was intentional, and the booster was towed back to port.

On July 14, 2014, the SpaceX team took a second shot at a controlled descent with landing legs after delivering an ORBCOMM satellite into space on the second-stage rocket. After touching down softly on the water, the rocket fell sideways into its final water "safing" state (a near-horizontal position that allowed tugboats to return it to port). But, sadly, and despite another nominal drop under engine power, the "water impact caused loss of hull integrity," causing the vehicle to sink to the depths of the ocean, according to a statement from the company. However, the firm had already received all data from the operational flight and planned to correct the error that caused the loss of the booster. The firm was also satisfied this test had confirmed that "the Falcon 9 booster [can] consistently" reenter the atmosphere from space "at hypersonic velocity, restart main engines twice,

Falcon 9 lifts off with ORBCOMM-2 from SpaceX's launchpad at Cape Canaveral.

deploy landing legs and touch down at near zero velocity."

The next launch of a CRS-4 satellite didn't include landing legs, since the much higher delivery altitude (to the International Space Station [ISS]) required too much fuel to guarantee a worthwhile landing attempt on descent. But on September 21, 2014, the first-stage booster did manage a controlled descent before toppling onto its side. One of the strangest features of SpaceX's reusable rocket program is the reliance on autonomous drone ships to "catch" the first-stage Falcon 9 boosters. But space programs don't happen in a vacuum, and Musk's idea was so crucial to reusable rocket systems that it seeded a legal dispute between Blue Origin and SpaceX in 2014, both arguing dibs on the sea-based retrieval concept.

Blue Origin already held a 2010 patent on the sea-based landing platform, but SpaceX was leaps and bounds beyond Bezos's technology. In the Blue Origin patent image, the similarity to Musk's drone ships is striking. But, SpaceX ultimately beat Blue Origin to the punch of actually building drone ships in the latter half of 2014. One was moved into each of the major oceans, and both were named after artificially intelligent and planet-sized spaceships in a favored science fiction series of Musk's called *The Player of Games*. Their names, which will forever be difficult to type in a flash, are Just Read the Instructions and Of Course I Still Love You, the former stationed in the Pacific, the latter in the Atlantic.

SpaceX's first attempt to successfully land a Falcon 9 rocket on a drone ship happened on January 10, 2015. After lifting the CRS-5 into orbit, the mission ended in a spectacular and fiery explosion after the vehicle's hydraulic systems that controlled the grid fins exhausted their fluids before touchdown. Lacking a means to stabilize itself, the rocket swerved away from its target and shattered into countless pieces. It wasn't until April 14 of that year that the firm had another go at a drone ship landing. After delivering the CRS-6 into a viable ISS rendezvous trajectory, the first-stage booster made a smooth, almost perfect descent—except for a slow throttle valve that kept the exhaust propellant firing for too long, adding extra lateral motion to the vehicle. Because of this sticky throttle, it calmly fell to its side before undergoing what Elon Musk would come to call a

← Falcon 9 is seen at Vandenberg Air Force Base Space Launch Complex 4 East with the Jason-3 spacecraft onboard on January 16, 2016. Jason-3 is an international mission led by the National Oceanic and Atmospheric Administration.

→ SpaceX's Grasshopper test vehicle photographed at the SpaceX Rocket Development and Test Facility in McGregor, Texas

"rapid unscheduled disassembly," his euphemism for "it exploded." When it comes to expected failures, at least, Musk isn't without a sense of humor. But that doesn't mean he and his company don't take time to assess serious problems with a sober mindset.

For example, in June 2015, a single strut that tethered a helium bottle inside a Falcon 9 broke midflight, triggering a chain reaction. "One of the struts failed and was unable to hold the helium bottle down, and the helium bottle would have shot to the top of the tank at high speed," said Musk in a call with reporters the following month. When the helium bottle struck the top of the tank, it created a colossal explosion that ended the mission, destroying the entire vehicle and its payload. For six months, the company worked to correct the issue, and then, in late December 2015, it returned to operational service, launching satellites for the company ORBCOMM. The payload was delivered sixty-two miles (100 km) above the surface of Earth (traveling at the speed of sound) before the rocket descended for the company's first-ever landing attempt not at sea but on land.

And, incredibly, the rocket made a soft, successful touchdown in the darkness of night, marking the first time in history that a VTOL rocket had lofted a satellite into orbit and then immediately descended for a safe return to Earth. It's hard to overstate the significance of this: SpaceX had officially engineered an operational launch system that could deliver a payload to space and return to Earth, vertically and under its own power and autonomous guidance, for a safe landing. Once again, what was once resigned to the pages and screens of sci-fi novels and films was happening before our very eyes.

Riding on this success, SpaceX aimed for another landing weeks later, on January 7, 2016. This time, after deploying its payload called Jason-3, Musk's aerospace firm would attempt to land on an autonomous drone ship. But, after its landing gear touched down on the pad and in what seemed like a random act of cruel chance, one of the legs didn't lock into place, causing the entire rocket to fall to its side and explode like its predecessors. Several meme videos had begun to flood social media, showcasing the long line of explosive endings to SpaceX's ill-fated landing attempts. But Musk and his firm showed a willingness to play its

"Well, technically, it did land . . . just not in one piece."

—Elon Musk

Elon Musk, left, shakes the hand of SES CTO Martin Halliwell after a Falcon 9 rocket, powered by a previously flown first-stage rocket, blasted off from Kennedy Space Center on March 30, 2017.

vulnerabilities as strengths, releasing its own supercut of every failed landing attempt titled "How Not to Land an Orbital Rocket Booster."

During one of the "rapid unscheduled disassembly" clips, a caption displayed under the explosion read: "Well, technically, it did land . . . just not in one piece." In March 2016, a Falcon 9 delivered a satellite into geosynchronous orbit: Essentially the satellite would maintain its station relative to the geographic location below. The payload was SES-9, slated to orbit 22,000 miles (35,406 km) high. But, while SpaceX had done this before, it had never done so with a landing in mind. This is especially difficult because, since geostationary satellites need to be inserted into orbits several tens of thousands of miles higher than most others, the amount of energy (and thus fuel) required for ascent is many times higher. This leaves less fuel for the subsequent descent, and in the case of the March 2016 launch, the Falcon 9 rocket hadn't slowed itself to safe velocities and smashed into a fiery end upon attempting touchdown.

But the very next month, on April 8, 2016, Elon Musk's firm finally did it. After lifting CRS-8 into orbit, a Falcon 9 booster descended out of a deep-blue sky, reignited its engines, and made a perfect, graceful landing on a drone ship, only a few feet (1 m) from the center of the target. SpaceX shared a special 360-degree video on its YouTube channel, enabling fans, enthusiasts, and whomever else to experience the ringside view for this historic first. In May, the firm did it again, ramping up the challenge by landing a Falcon 9 after delivering a satellite to that 20,000-mile-high (32,187 km) geostationary orbit (GTO). Even without the extra fuel, the drone-ship landing was nearly perfect. Another video showed that the touchdown was centered almost exactly on target.

For good measure, SpaceX performed the same feat, nailing the drone-ship landing of another Falcon 9 and the second GTO mission to do so. It wasn't a fluke: The math, along with the autonomous software guiding the booster down, checked out.

But this doesn't mean things can't go wrong. In June, one of the Merlin engines put out a low level of thrust, which made for another messy ending. But in July 2016, this hiccup was corrected when the second-ever return descent to the same launchpad in Cape Canaveral ended in another graceful landing, following a quick flight to the International Space Station.

However, Musk and his private aerospace firm weren't interested in becoming a smaller, cheaper version of NASA. With aspirations of building a new means of transporting ordinary people via ballistic flights through space, he had to test the portability of the reusable flight system. That meant going beyond Cape Canaveral and the familiar safety of the Atlantic. So, in early January 2017, SpaceX launched a Falcon 9 from Vandenberg Air Force Base in California, which then was successfully retrieved from the Pacific Ocean. After accomplishing nearly every task in the design of a reusable rocket system, from a less expensive launch system to the drone landing and retrieval of the first stage, it was finally time for SpaceX, and Elon Musk, to take the final step in proving the worth of a reusable alternative to federal space agencies like NASA. A used rocket needed to launch again.

On March 30, 2017, a Falcon 9 that had previously lifted the CRS-9 mission for NASA was refurbished and equipped with another satellite, this time from the European space communications juggernaut SES. Without a single recorded error, the vehicle lifted the payload to its orbital trajectory and then descended autonomously under its own power and made a smooth landing.

In the coming years, this success became routine for SpaceX, as it lifted higher and higher-mass payloads into orbit, reusing boosters for twenty of its missions. Reusability was no longer a concept to prove—it had become a habit, a new tool with which Elon Musk could begin to take serious strides in the space industry like no private company had ever done before.

5 SUSTAINABILITY ON COSMIC SCALES

In the days of Andrew Carnagie, J. P. Morgan, and the other robber barons, the economy centered on steel, oil, and railroads. The only common dangers to sea life in the nineteenth century were whalers, who were popular (*Moby-Dick* wasn't). And rather than reinventing themselves under global initiatives to build smart cities with sustainable infrastructures, American cities like New York were racing to catch up with the industrial might of London.

All to say that, within the shared span of the space barons' lives, nearly every U.S. manufacturing industry was replaced by service industries. New York became the undisputed epicenter of a global finance economy, and the whales became so endangered that pop icons like Michael Jackson and sci-fi film franchises like *Star Trek* raised awareness on the issue. Initially, the Internet offered what many believed would become a decentralized public commons, but then was bought and divided by major corporations into social media interfaces that spied on us and flattened our social values as citizens of Facebook, then Twitter, Instagram, TikTok, and so on.

The newly deployed iROSA solar panel array, equipped on the ISS

Meanwhile, Ray Kurzweil predicted "the technological singularity" on the cultural horizon as the ravages of climate change became absolutely undeniable. All of these factors and more shifted the scatterplot of innovation toward very different meters of success than the ones enjoyed by the nineteenth-century robber barons—ones that could expand the reach of the space barons much farther than their predecessors as they aligned their commercial empires with global policies increasingly centered on the concept of sustainability.

Sustainability in Space—from Sputnik to Starlink

According to the UN World Commission on Environment and Development, "sustainable development is development that meets the needs of the present without compromising the ability of future generations to meet their own needs." Other definitions in vogue in academic environs stress a need to preserve social equity, resilient communities, and an informed grasp of how the world as we know it (which is to say as a habitable planet) is interconnected. A habitable planet can continue only if humans in their role as stewards of the future apply a systems approach that affirms the inherent complexity of life and its manifold relationship to the planet.

As entrepreneurs purporting to reimagine the future of humanity in space, the twenty-first-century space barons needed to reassess the material legacy of the previous Space Race that extends far beyond its end in the 1970s and 1980s. Considered as a twenty-first-century business, space programs of the last sixty years were a straggling industry not overwhelmingly concerned with what, beyond clear-cut legacy successes like the moon landings, it left behind. Much of this was a consequence of Congress's increasingly scant funding to NASA (comparably, and adjusted for inflation, of course).

Among the negative effects of orbital space missions, most portentous is the looming danger of space junk—the mountain of defunct or damaged satellites in low-Earth orbit (LEO). Since the late 1950s, satellites that weren't intentionally deorbited to burn up in the atmosphere could end up joining a very dispersed but maximally hazardous bulk of space junk. Flying at mind-shattering speeds of 17,000 mph (27,359 km/h), space junk can damage or destroy other operating satellites, or

← Space junk rapidly grew into a serious problem in the 2010s, crowding low-Earth orbit.

← Modern cities are transforming into smart cities, which emphasize a new value of "resilience" and a higher integration into the Internet of Things (IoT).

↓ Some have posited that the Earth is not only interconnected but also a living, breathing organism.

↓ A hilltop transformed into a solar power–generating plant

even crewed missions. According to a report from The Conversation, roughly 7,941 satellites were in orbit as of mid-September 2021.

Since the now-defunct Soviet Union launched Sputnik, the first artificial satellite, in 1957, space programs have continued to lift satellites into orbit. Through the latter half of the twentieth century, the number of satellites grew at a steadily increasing rate, rising from a rate of sixty to one hundred or more every year until the early 2010s—which brings us back to Elon Musk's SpaceX. Starlink, the tech billionaire's bid to provide internet to every underserved corner of Earth, saw an unprecedented rise in the number of satellites launched. From 2010 to late 2020, 114 rocket launches had carried 1,300 satellites into space, breaking the previous record of 1,000 new satellites per year. But in 2021, this record was again broken, with more than 1,400 satellites launched into orbit.

This exponential growth owes much to the way SpaceX lowered the demands and cost for lifting mass into orbit. But advances in computer and electronics technology have also enabled the creation of smaller satellites, which means less mass and more room on each launch. An overwhelming 94 percent of the spacecraft launched in 2020 were "SmallSats," each weighing less than 1,320 pounds (600 kg). Not content to miss a chance to claim part of a nascent $1 trillion industry, Bezos and Amazon are also developing an internet satellite network, called Project Kuiper.

Today, it's no exaggeration to say we live under a crowded sky. In the wake of SpaceX's launch of its initial sixty Starlink satellites, astronomers worldwide complained that Musk's low-Earth orbit constellation blocks starlight. And this cosmic blinder extends beyond the visible spectrum of casual stargazing: Radio astronomers might lose 70 percent sensitivity in critical frequencies because of Starlink's overwhelming mega-constellation, reducing our ability to study the very early universe, distant galaxies, and even signals from potentially inhabited planets beyond our solar system (at least from Earth's surface). This rapid increase in the number of satellites also makes orbital collisions more likely, and when two

bodies of metal and silicon collide at supersonic speeds, they can send space debris shooting around the planet at unconscionable velocities.

In reducing the costs of space-launch systems while increasing the efficiency of orbital delivery profiles, Musk and Bezos are also revealing an ironic feature of the ongoing expansion into space: Even hundreds of miles above Earth, industries designed to enhance the life dwelling below it come with a catch. Consequently, the space barons need to account for and minimize the impact of continued growth on other human endeavors. For its part, SpaceX already tested a handful of scenarios to lower the frequency of Starlink collisions.

Meanwhile, Amazon has released plans to intentionally burn up their satellites via deorbit no more than 355 days post-completion of missions.

A great many stakeholders in space depend on finding a balance that won't throw a wrench in ambitions for commerce, experimental missions, and other human aims. In this sense, "sustainable" implies keeping in mind not only the mission itself but also how to control the way it ends to ensure that future missions and generations will benefit, not suffer, from the advances of the second Space Race.

The Shortcomings of Sustainable Technology

On Earth, the same principles ostensibly motivate advancements in novel power sources, like wind and solar. Musk is no stranger to solar power, having directed his all-electric automaker company Tesla to develop a suite of state-of-the-art solar-cell products. Each Starlink satellite uses a flat solar panel that weighs 573 pounds (260 kg). And since they're launched in a stack, the internet satellites can self-deploy. They also use krypton-fueled thrusters for propulsion, in addition to orbit modifications and maintenance, and eventual deorbiting maneuvers.

As another backup to avoid becoming space junk, the satellites can autonomously alter their own trajectory, which is detected and plotted with uplinked tracking data. When they've reached the end of their life span, the satellites perform a deorbit burn, which eliminates 95 percent of the material according to *National Geographic*. But compared with other solar implementations, commercial solar power for homes and even Starlink's solar technology are modest. In June 2021, NASA began installing new solar arrays jointly designed by Boeing and Redwire on the International Space

← Starlink satellites have contributed to the sudden increase in the number of low-Earth orbiting objects. Modern astronomers are rightly worried that Starlink satellites could interfere with Earth-based observations of the universe.

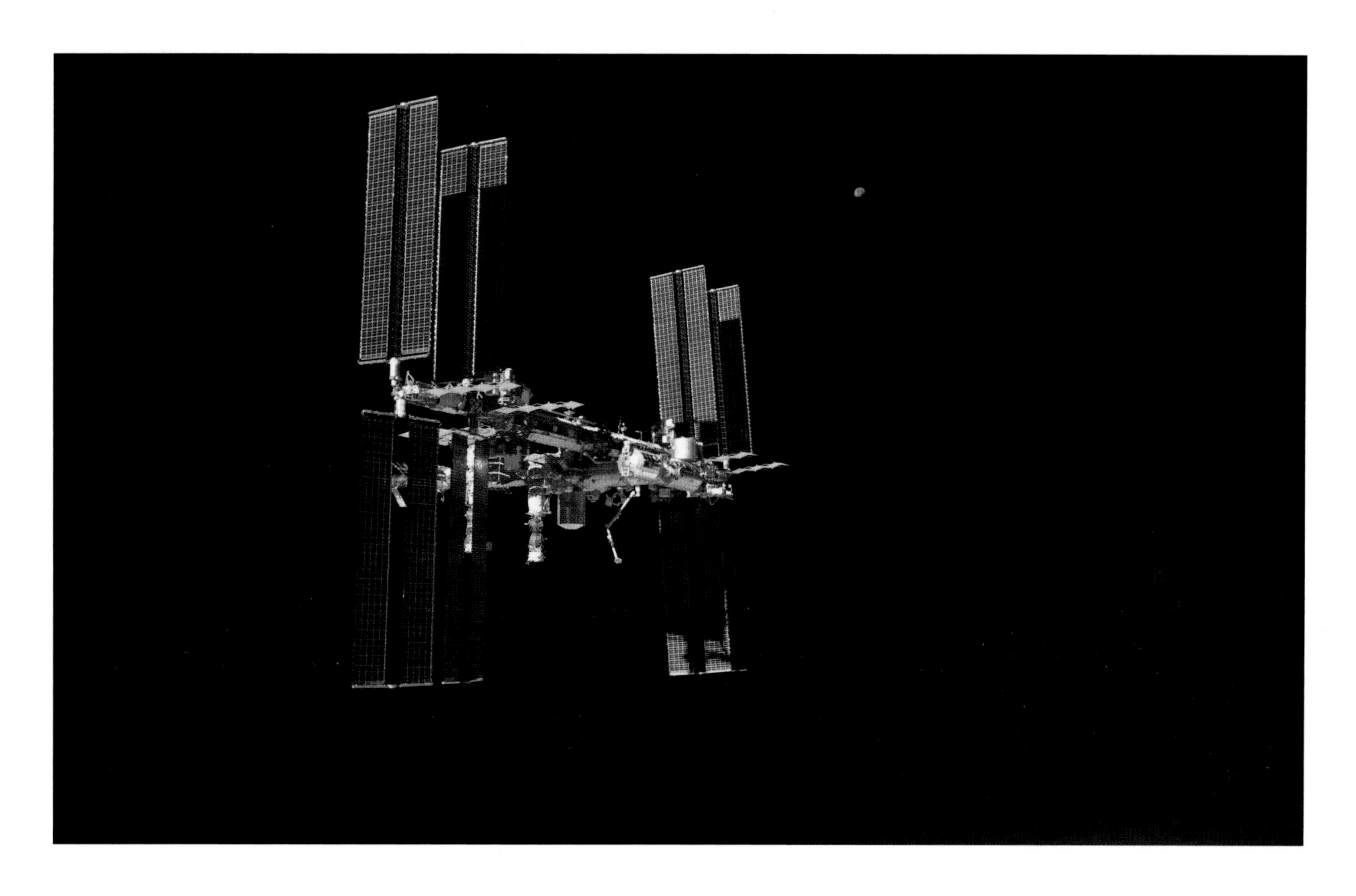

Station. Dubbed the International Space Station Roll-Out Solar Arrays (iROSAs), they were manufactured in Boeing's Spectrolab, and each one generates more than 20 kW (27 hp), for a sum of 120 kW (163 hp) for the entire station. This increases the ISS's maximum power capacity by 20 to 30 percent and (as of 2021) was slated for completion no earlier than 2023. But innovations in LEO can only go so far, especially when the most symbolic structure of the interim years between the first and second Space Race, the ISS, is nearing the end of its days.

In a way, the available history of Space Race 2.0 is only the beginning. We're just not that far yet. And if, in this case, the past is prologue, then the first step was developing a reusable launch system that can function without relying on NASA and other twentieth-century space agencies. But the second step? That's our return to the moon, this time to stay. It's a tall order, with the potential to outrival the darkest moments of the Apollo program. Indeed, Musk's 2016 warning that an eventual extinction event on Earth necessitates the settlement of at least one more world changes the meaning of space travel from a noble trek of exploration into a desperate attempt to preserve humankind.

But behind the scary motivations of survival and the fear of extinction is something more relatable and modern: the ever-evolving and accelerating advocation of sustainable practices. In the early 2020s, with wildfires spreading smoke across two continents, increasingly costly extreme weather

↑ The International Space Station could be gone before the end of the 2020s.

↓ California wildfires have expanded in scope and severity to unprecedented levels.

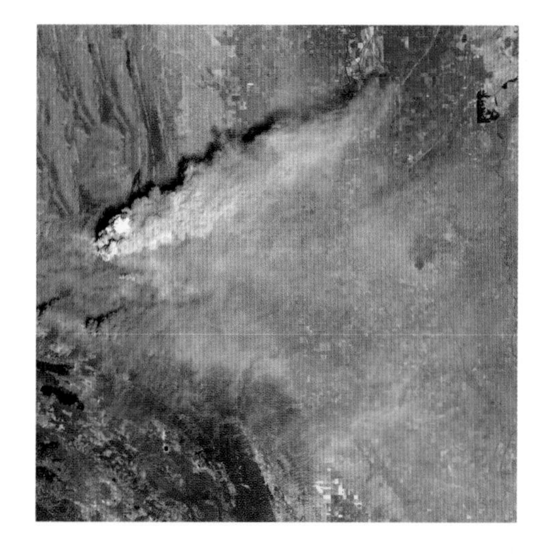

events, rising average temperatures, and even industrial shortages of semiconductors, we on Earth are living in a very strange time, when corporate PR campaigns are forced to overlap with real, verifiable science. Corporations in nearly every industry, from Ford and Tesla to fossil fuel heavy-hitters like British Petroleum (BP), have spent the last decade in a quick pivot to accommodate sustainability as the new and most central value of the human condition.

Of course, you don't have to take major corporations at their word. When major automakers share carbon-zero goals that align with the Paris Agreement, that's more than PR. But other times, ostensibly sustainable practices on the part of major firms are designed to disguise and obfuscate a failure to comply. For example, an August 2021 op-ed in the *Guardian* by renowned author Rebecca Solnit revealed that the concept of "carbon footprint," a catchall buzzword for measuring personal impact on the natural environment, is actually a marketing phrase coined by an advertising company hired by BP.

The goal was to make average citizens blame themselves and their lifestyles for climate change so that the general perception would shift responsibility from the central role that fossil fuel firms play in generating pollution and ecological calamity. In this negative sense, sustainability would have less to do with ensuring human prosperity than with lining the pockets of C-level officers of harmful industries, despite chunks of ice shelves bigger than the state of Rhode Island breaking off the Antarctic continent.

The appeal of electric vehicles (EVs) is obvious at first glance: Instead of pumping toxic fumes into the air with every transit to and from work, you can become part of the solution and rest assured in the knowledge that you're no longer part of the problem. But the mining of lithium for the lithium-ion batteries used in EVs is far from sustainable, in the sense of having zero negative effects on the environment.

← Lithium battery packs on an EV assembly line. Electric vehicles don't do much for the environment if the industrial complex behind them is just as bad or worse than that used to extract fossil fuel.

→ Lithium mines form artificial bodies of mineral-heavy brine that evaporate.

More than half of the lithium used for EVs is mined from the "Lithium Triangle" region of Bolivia, Argentina, and Chile. Miners drill holes in the salt flats and pump a salty, mineral-heavy brine up to the surface, where it forms artificial bodies of water that evaporate.

This process wastes approximately 500,000 gallons (nearly 2 million liters) of water for every ton of lithium mined, which has a drastic effect on the livelihood of nearby farmers. Toxic substances can get into the water supply, which happened in Tibet, where hydrochloric acid found its way into natural bodies of water and killed massive amounts of aquatic life. EV batteries also need cobalt and nickel, the former of which is often extracted from mines that don't shy away from engaging in child labor in extremely dangerous settings. Cobalt mining generates vast quantities of airborne particulates, often uranium, the same radioactive element used to bomb Hiroshima and Nagasaki.

And, as of the early 2020s, any environmentally harmful practices carried out by Western industry are almost always more harmful as practiced in China. According to a *Forbes* report, the production of EV batteries in China produces roughly 60 percent more CO_2 pollution than combustion engines. Safe to say that, on Earth, "sustainable" practices can turn out to be almost the exact opposite.

Preserving the Moon and Mars

So what will sustainable practices mean in space? Commercial expansion conducted in environments that are already naturally deadly to humans provides an opportunity to practice a kind of sustainability stripped of its need for public relations, at least in the often-irradiated confines of its space-based destinations. Despite the space junk cluttering the upper atmosphere, no one in Earth space has territorial claims on one or another region. Space colonies are still a figment of collective imagination, in sci-fi films, shows, and manga.

The only sensible definition of sustainability on alien worlds is one that enables humans to thrive using only the resources immediately available in the surrounding environment. You can't make the moon any less habitable than it already is. There's practically no atmosphere, and even a nuclear bomb would be comparably muted since there's no gas to propagate a sound or shockwave. In other words, at least on the moon, there's almost certainly no biosphere to save. But that doesn't mean we have no responsibility to preserve the natural conditions of the moon or other alien worlds.

Within nuclear reactors, uranium and other fissile material is actually very safe. But cobalt mining can release the radioactive element into the air.

Lacking proper atmospheres, space and the moon are extremely deadly and radioactive environments, which means our thin atmosphere is a tiny bubble of habitability. Few understand this reality.

There may be resources like water and essential minerals on the moon that can sustain a human presence. But the moon, like Earth, cannot support infinite industrial growth. Additionally, no one can technically own the moon, since the moon "and its natural resources are the common heritage of [humankind]," according to the Moon Agreement adopted by the UN's General Assembly in 1979, which entered into force in July 1984. The agreement also states that the moon and other cosmic bodies "should be used exclusively for peaceful purposes, that their environments should not be disrupted."

"[A]n international regime should be established to govern the exploitation of such resources when such exploitation is about to become feasible," read the UN agreement. This sounds like a good idea, but major powers and corporations might choose to ignore this international consensus (like they often do down here) and exploit the moon's resources according to their own interests.

Beyond self-sustainability on the moon's surface, one should at least hope that the greatest priority will be to preserve environments long enough for thorough and exhaustive scientific exploration, investigation, and analysis. The origin story of Earth, the solar system, and even the way other alien worlds may form far, far away is written in the worlds around us. There probably isn't life on the moon or Mars (at least, nothing approaching the complexity found on Earth), but on the Red Planet, we might find solid evidence of past life. So, if there's no stopping human industry from going beyond

← Many alien worlds, including Mars, have no known biosphere, which means sustainability must take on a different meaning. But this does not mean we have the right to damage or bespoil them.

→ According to the UN, the moon belongs to no one.

"The moon and other cosmic bodies should be used exclusively for peaceful purposes, that their environments should not be disrupted."

—UN General Assembly

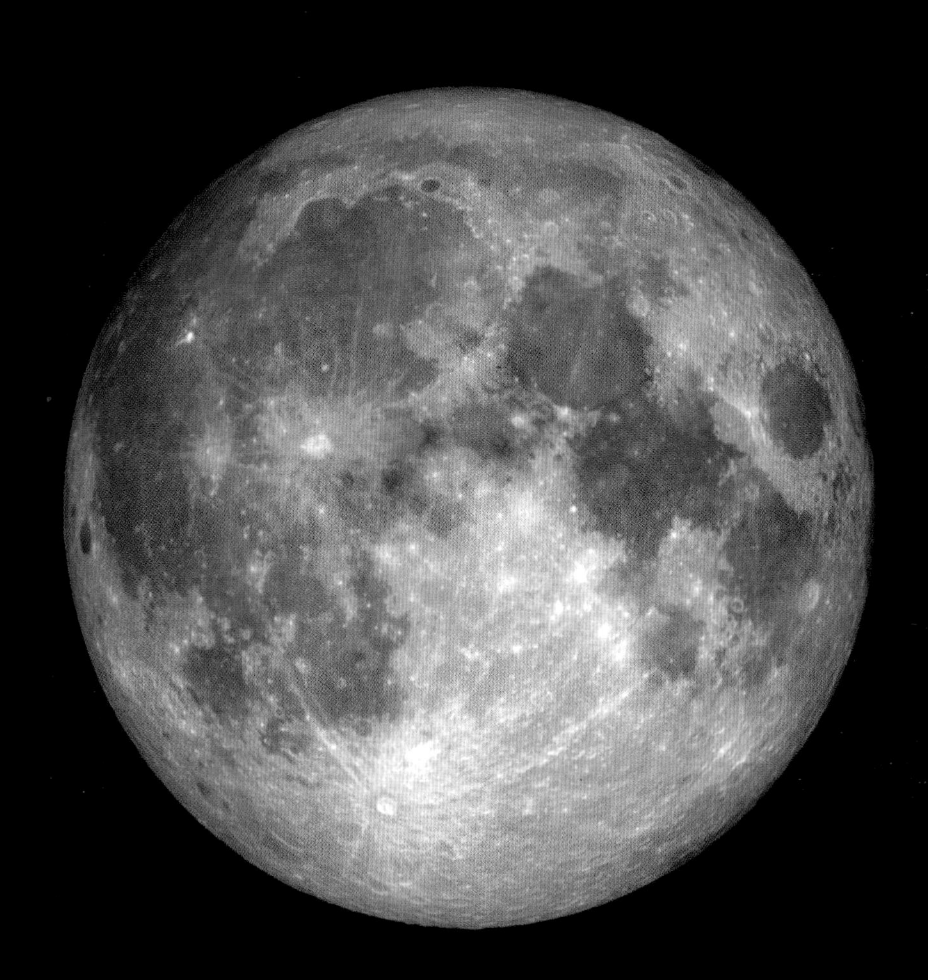

self-sustainability like it often does on Earth, we should try not to jettison this minimally noble goal of not destroying the evidence of where we come from, what else is out there, and how the universe works for the comparably short-term goals of profit, luxury, and cosmic empire that likely motivate the space barons (for example, Musk changing his title at Tesla to "Technoking").

You could say the long-term settlement of humans on the moon and Mars hinges on developing a reliable system of self-sustainability. But there's no guarantee that the major powers, space agencies, and aerospace firms of Earth will stop here, even out of respect for the very empirical science that will have taken them to space. In other words, we have a long way to go before we achieve even the most conservative level of self-sustainability worthy of interplanetary expansion. But there's no shortage of ideas about how to get there, keep ourselves alive once there, and build local resource-extraction processes to extend human economy into space.

Building Interplanetary Industry

Early in 2021, the Defense Advanced Research Projects Agency (DARPA) kicked off a new program called the Novel Orbital and Moon Manufacturing, Materials and Mass-efficient Design (NOM4D). Its aim is to pioneer new methods of manufacturing and production "off-world" that can support large-scale structures, both in space and on the lunar surface.

In short, the U.S. military industrial complex wants to build factories on the moon. "Manufacturing off-Earth maximizes mass efficiency and at the same time could serve to enhance

↑ The planet Mars may have once harbored indigenous life, whose traces may await our discovery.

→ A conceptual illustration of a future base on the moon

stability, agility, and adaptability for a variety of space systems," said Program Manager Bill Carter of DARPA's Defense Sciences Office, in a report from the agency. Of course, this isn't a new idea. In the initial years of the first Space Race (1959), the U.S. Army began designing a program to build a physical military base on the moon by 1965. Of course, this never got off the ground. Back in the hottest days of the Cold War, and with comparably primitive technology, it just wasn't in the cards. But now, NASA's international private and public collaborative program for returning to the moon, called Artemis, is quickly becoming the accelerating force for a new and sustainable infrastructure in space. To accomplish this, astronauts will require a reliable form of sustainable energy to power lunar habitats, robotic and crewed rovers, and complex construction systems.

In March 2021, NASA selected five commercial companies to develop novel solar array systems on the moon that deploy vertically, springing open in an upward direction to stretch thirty-two feet (10 m) above the lunar surface. The companies selected by NASA are Northrop Grumman Space Systems (NGIS), Lockheed Martin, Maxar Technologies' Space Systems Loral (SSL), Honeybee Robotics, and Astrobotic Technology. "Having reliable power sources on the moon is key to almost anything we do on the surface," said Niki Werkheiser, director of Technology Maturation at NASA's Space Technology Mission directorate

PROJECT HORIZON
Volume I
SUMMARY AND
SUPPORTING
CONSIDERATIONS
URANUS
PLUTO
MARS
VENUS
SATURN
SUN
MERCURY
EARTH
MOON
JUPITER
UNITED
STATES
ARMY

↑ NASA is adapting solar technology into vertical configurations for use on the lunar surface.

← A depiction of the U.S. Army's defunct plans for a moon base, called Project Horizon

(STMD), in a NASA release. "By working with five different companies to design these prototype systems, we are effectively mitigating the risk that is inherent in developing such cutting-edge technologies."

Conventional solar arrays and deployment structures in space are designed for operation in microgravity environments or on flat, horizontal surfaces. The novel vertical arrangement will enable the moon-based solar array to receive sunlight more effectively, since the sun doesn't rise high in the lunar sky when you're based on the south pole. This is exacerbated by the great abundance of hills, slopes, and rocky formations that can cast ink-black shadows across the surface, nullifying the effectiveness of flat, horizontal solar arrays. "These solar power designs could help enable continuous power for Artemis lunar habitats and operations, even in areas that are shaded by rocky features," said Lead in Vertical Solar Array Development Chuck Taylor of NASA's Langley Research Center in Hampton, Virginia, in the NASA blog post. Taylor added that the ongoing pursuit of more efficient solar arrays will reveal unforeseen applications on Earth. For example, private homeowners and business owners could upgrade their own solar array efficiency with reliable, vertically oriented designs since these could reach above the shade of nearby trees or taller structures.

Space-based solar power (SBSP) has entered a critical threshold of becoming a viable alternative to power the entire planet with sustainable energy. Some proposals have theorized the construction of colossal mirrorlike reflectors positioned in orbit that can concentrate the sun's light, convert it into electromagnetic radiation, and beam it down to the surface with a laser or in the form of microwaves. Similar efforts might one day power human settlements on the moon. But this would be far easier if we could mine the resources for constructing sustainable power-generation infrastructure (and the assembly facilities themselves) from the moon itself.

NASA and Private Firms

In June 2021, NASA administrator Bill Nelson addressed the House of Representatives to, among other things, correct a common misconception about the first Space Race: the idea that NASA designed, built, and executed all space missions in feats of taxpayer-funded ingenuity. It turns out, this is only two-thirds right. "In the Apollo Program," Nelson began, "we got to the moon with American corporations." At least a dozen major U.S. firms helped NASA achieve the world-historic success of placing humans on the surface of the moon. Scientists and engineers at the agency designed the mission profile and developed much of the technology required to make it happen. But the physical production of space capsules, rockets, the moon lander, spacesuits, moon buggies, and rovers happened in collaboration with the most advanced technology companies in the country.

Apollo was a publicly funded, federal space program. But it wouldn't have gotten off the ground without substantial contributions from the private sector. This is significant because it's relevant to Space Race 2.0, which in the last decade has seen the reorientation of space programs around novel flight systems like reusable rockets, specifically from the private sector. Instead of doing all the engineering in-house on new space vehicles to loft astronauts and other cargo to the International Space Station, and then paying a major firm to build it, NASA now outsources the whole process.

With increasing regularity, the U.S. space agency announces calls for design proposals that can meet the needs of mission profiles, and the winning bidders do more than build the spacecraft: They also operate them, with light to moderate supervision from NASA. This new mission-development flow laid the groundwork for SpaceX's rise to power in the new age of private aerospace firms. And, there are several features to this new partnership.

First, all the engineering goals, including the concept of reusability, were already explored and attempted before with varying degrees of success (cf., the Space Shuttle). This means private aerospace firms weren't exactly starting from scratch, and NASA knew what to look for. Private space companies can also invest their own money, which, on paper, can save public taxpayer money. Although, if the federal government subsidizes these firms and their development, the line between public and private money is more blurred. These firms also take on more of the risk than NASA and can apply the cold efficiency of twenty-first-century corporate decision-making structures and hierarchies, instead of the oft-stymied bureaucracy of government programs, which are forever tethered to money from Congress. Private space firms don't have to constrain their programs into neat four- to eight-year packages that can be designed and implemented during a single presidential administration for fear of seeing a program axed in the next one.

However, not everyone approves of the new public-private partnership system, which grants more authority and control than ever before to private

↑ Bill Nelson, NASA administrator in the early 2020s, argues that space travel and corporate compliance are more than compatible. To him, they are symbionts.

→ Without corporate partnerships, Apollo 11's historic journey to the moon may never have happened.

UNITED STATES
USA
USA

The Space Shuttle Endeavor, now retired, is docking with the International Space Station. These were among the first attempts at enhancing the sustainability of spaceflight by using a wholly reusable vehicle.

companies. NASA is paying space barons and their firms to develop, build, and operate robotic probes to explore the moon, in addition to the life-support systems needed to sustain humans on deep-space missions and the next-gen landers that will finally return humans to the lunar surface. Some of the House lawmakers in the June 2021 presentation, like Science Committee chair Eddie Bernice Johnson, expressed concerns that private aerospace firms should assume all the risks associated with these ventures. She's not entirely wrong.

The Lunar Gateway

Low-Earth orbit begins at about 60 miles (97 km) high, with the International Space Station maintaining altitude at roughly 220 miles (354 km). The moon is 238,900 miles (384,472 km) away. That's nearly 1,000 times as far and a lot more time and distance for something to go wrong, without even mentioning the challenges posed by landing a vehicle on another body in space. Even NASA had close calls in lunar transit, most notably during the Apollo 13 mission. But the agency, along with new and old private partners like SpaceX, Blue Origin, Lockheed Martin, Boeing, Northrop Grumman, and more, are going to the moon. And they already know how it's going to happen.

A crucial stepping-stone for NASA to return humans to the moon is the need to establish a permanent space station in lunar orbit. Called the Lunar Gateway, it will serve as a critical docking hub for astronauts to make a "pit stop" on their way to the surface and beyond. It will look a lot like the ISS, but smaller.

↑ In the 2010s, NASA began implementing a new strategy that placed private firms in roles of much greater control than ever before.

→ This illustration forces perspective—the moon is nearly 1,000 times as far from Earth as the International Space Station in the foreground.

Engineers at the Space Mission Analysis Branch (SMAB) of the Langley Research Center have led research on developing the Gateway, but engineers at NASA's Kennedy Space Center are playing a more central role. The Lunar Gateway's function will be to operate like a mini-ISS that can dock with the Orion spacecraft, NASA's next-gen spacecraft for crewed missions into deep space, under development by Lockheed Martin.

The Lunar Gateway will incorporate an advanced modular design, enabling the implementation and integration of varying parts can launch and assemble separately. The Gateway is also slated to last at least fifteen years, during which time it will be continually modified and upgraded to support evolving mission objectives. As of May 2020, the Gateway was scheduled to begin construction by 2023, which would have seen the power unit and the Habitation and Logistics Outpost (HALO) unit in lunar orbit, already capable of sheltering astronauts for short-term stays. If all went according to plan, the Gateway could have supported astronauts for months at a time by 2025.

The Gateway is conceptually one part of a two-pronged campaign launched by former vice president Mike Pence, who instructed NASA to achieve the capability to maintain a sustained human presence on the moon by 2028. In June 2021, it seemed like NASA was on schedule to meet its 2023 deadline for lunar orbital insertion of the Gateway when it finalized a contract to develop the HALO with Northrop Grumman, which won a firm, fixed-price contract of $935 million.

↑ Robert Zubrin, who was among the first to advocate for a more sustainable mission architecture for a mission to the Red Planet (called Mars Direct and later, Mars Indirect), presented his case for Mars to a Senate Committee in 2003. Zubrin, a former engineer from Lockheed Martin, was also a mentor to Musk.

← The Habitation and Logistics Outpost component of the Gateway, a critical part of NASA's Artemis program

→ The Lunar Gateway will serve as a "pit stop" for astronauts on their way to the moon and beyond.

"Under the contract, Northrop Grumman will be responsible for attaching and testing the integrated HALO with the Power and Propulsion Element (PPE, being built by Maxar Technologies)," read a blog post from NASA's official website. "Northrop Grumman will also lead the integrated PPE and HALO spacecraft turnover and launch preparation with SpaceX, and support activation and checkout of HALO during the flight to lunar orbit," it continued, referencing Elon Musk's firm, which was the sole winner of a contract with the agency to design and operate a crewed lunar lander: the forthcoming Starship vehicle.

"NASA is building the infrastructure to expand human exploration [farther] out into the solar system than ever before, including Gateway, the lunar space station that will help us make inspirational scientific discoveries at and around the moon," said Bill Nelson, NASA administrator, in a blog post from the agency. "Just as importantly, these investments will help NASA carry out the United States' horizon goal: to further develop and test the technology and science needed for a human trip to Mars." Northrop Grumman's HALO design is closely based on the firm's Cygnus spacecraft, which had already made fifteen successful resupply missions to the ISS when it won the Lunar Gateway HALO contract.

Another private aerospace partner of NASA, Lockheed Martin, also made waves in the forthcoming Artemis program when it finished assembly of the Orion Artemis I spacecraft in January 2021, after which custody was transferred to NASA's Exploration Ground Systems (EGS) team. Once it launches, Orion will become NASA's new flagship exploration-class spaceship and part of the agency's Space Launch System (SLS) rocket. Before humans make the trip, Orion will make an uncrewed flight to the lunar orbit and back to Earth in a test and round-trip validation mission that will take three weeks. "Orion is a unique and impressive spacecraft and the team did an outstanding job to get us to this day," said Orion's vice president and program manager, Mike Hawes, of Lockheed Martin, in a company press release. "The launch and flight of Artemis I will be an impressive sight, but more importantly it will confirm Orion is ready to safely carry humans to the moon and back home."

In September 2020, NASA shared an update and elaboration of the Artemis program's four stages. After

the initial two uncrewed flight tests of the Orion and SLS rocket to lunar orbit and back, Artemis III will carry astronauts to the moon, where, pending the final rollout of the Human Landing System (i.e., SpaceX's Starship), it will make a landing on the south pole. This will be the first crewed mission to the surface of the moon in more than half a century. Initially scheduled for 2024, two astronauts were to spend up to six and a half days on the lunar surface and perform at least two EVAs (extravehicular activities) before returning to orbit and reboarding the Orion spacecraft, which would take them back to Earth.

Initially scheduled for 2026, the Artemis IV mission was also slated for a crew, which would travel to the Lunar Gateway, delivering the integrated habitation modules (I-HAB) for incorporation into the larger Gateway system. There also exist plans for Artemis V through VIII, which will land more astronauts on the moon to build up infrastructure already *in situ* on the surface. Once on the surface, astronauts will bring a settlement into maturity, finalizing habitats, scientific instruments, rovers, and high-tech resource extraction equipment to begin the long-term project of making human presence on the moon a permanent, self-sustainable enterprise.

Incredibly, NASA and its commercial partners' leadership in designing reusable and sustainable systems to deploy on the lunar surface have inspired other industrial powerhouses to cultivate ambitions for deploying the most cutting-edge sustainable technology for lunar mobility. In September 2021, Honda Motor Company announced plans to assemble robots, eVTOLs (electric vertical takeoff and landing aircraft), and direct efforts to help settle the moon. Robots and eVTOLS have seen an accelerated success in recent years, but Honda's plans for the moon, although ambiguous, are particularly exciting. It's already joined forces with Japan's national space agency, the Japan Aerospace Exploration Agency (JAXA), to construct a "circulative renewable energy system on the lunar surface."

The success of such a system is contingent on the presence of water on the moon, and in October 2020, NASA discovered water on the sun-facing side of the moon, specifically in and around the Clavius Crater on the southern hemisphere. This precious substance is rather dispersed—on the scale of separate molecules too small to cohere into puddles of liquid or ice blocks. But with advanced hydrogen-extraction technology, NASA and its private aerospace partners could save billions of dollars simply by avoiding the daunting and expensive task of lifting water from Earth to the moon. "It's far easier to travel when you don't have to carry everything with you that you might need for the entire trip," said Chief Exploration Scientist Jacob Bleacher, of NASA's Human Exploration and Operations Mission Directorate.

Bezos's NASA Lawsuit and Moon Mission Delays

Initially, NASA had planned to place humans and the first woman on the moon by 2024, but these plans were scrapped with a whimper following an August 2021 audit that cast doubt on its original timeline. "[D]elays related to lunar lander development and the recently decided lander contract award bid protests will also preclude a 2024 landing," read NASA's audit. Elon Musk

After the Gateway is established, astronauts will begin the arduous task of constructing an extensive and permanent moon base.

replied to the news by tweeting that SpaceX would pick up the slack on developing moonwalk-capable spacesuit technology if NASA lacked the funds.

In late 2021, a November launch of Artemis I and the Orion spacecraft looked extremely unlikely, with a tentative next attempt slated for January 2022. But that, too, was further delayed. NASA's announcement was sad, but also frustrating to many was the implication that Jeff Bezos's Blue Origin, which had launched serious lawsuits against the agency for not being selected to build the Artemis landing system, had played a deciding role in delaying humanity's return to the moon. This came several months after NASA awarded SpaceX sole contractual rights to develop and operate the first commercial human lander for a return to the moon, a deal worth $2.9 billion.

In winning this contract, SpaceX bested not only go-to aerospace developer giants like Northrop Grumman, Lockheed Martin, and Draper, but also Musk's primary rival in the second Space Race: Bezos's Blue Origin.

"NASA's Apollo program captured the world's attention, demonstrated the power of America's vision and technology, and can-do spirit," said Lisa Watson-Morgan, the program manager for NASA's Human Landing System (HLS) officiating the development of the lander, in a press briefing. "And we expect Artemis will similarly inspire great achievements, innovation, and scientific discoveries. We're confident in NASA's partnership with SpaceX to help us achieve the Artemis mission."

This was a surprising outcome, not because SpaceX was a recipient of the lunar lander contract, but because NASA had originally declared there would be two private aerospace recipients for this program. Making SpaceX the sole recipient appeared to some like a move made specifically to exclude Jeff Bezos from Artemis HLS system development and a preferential move to fund Musk's Starship HLS (Human Landing Ship). Days after NASA's announcement, Blue Origin filed a lengthy fifty-page protest with the Government Accountability Office (GAO), according to an initial report from the *New York Times*.

"It's really atypical for NASA to make these kinds of errors," Blue Origin chief executive Bob Smith told the *New York Times*. "They're generally quite good at acquisition, especially its flagship missions like returning America to the surface of the moon. We felt that these errors needed to be addressed and remedied." It's a bit jarring to hear a private company refer to NASA like

The Orion Artemis I will become NASA's new flagship exploration class spaceship.

a doctor to a patient, but Musk had a different, more predictably risqué comment on the complaint. "Can't get it up (to orbit) lol," read his late-April tweet reply to the *New York Times* exclusive.

In July 2021, Bezos's aerospace firm wrote an open letter signed by the space baron, suggesting that NASA reconsider its decision to award SpaceX sole recipiency of the lunar lander project. To sweeten the deal, Bezos offered $2 billion as a payment waiver to the U.S. space agency. In his letter, Bezos explained that Blue Origin had formed a "National Team" of heavy-hitting tech firms, including Northrop Grumman, Lockheed Martin, Draper, and 200 other small-to-medium contributing companies headquartered in forty-seven states.

All these companies pooled resources and engineering know-how to design an HLS that was compatible with the full spectrum of launch vehicles on the table to return humans to the moon. Bezos reviewed NASA's reasons for initially asking for two separate lunar lander proposals: to lower the risk of mission mishaps, development delays, and excess cost hikes. While Bezos affirmed NASA's ongoing budgetary constraints, the language suggested that Blue Origin would simply fill in the financial gaps in the future, ostensibly adding more autonomy to NASA with regards to its federal funding, while implicitly making the agency's endeavors slightly more dependent on checks from the billionaire space baron.

Orion Artemis I was handed over from Lockheed Martin to NASA.

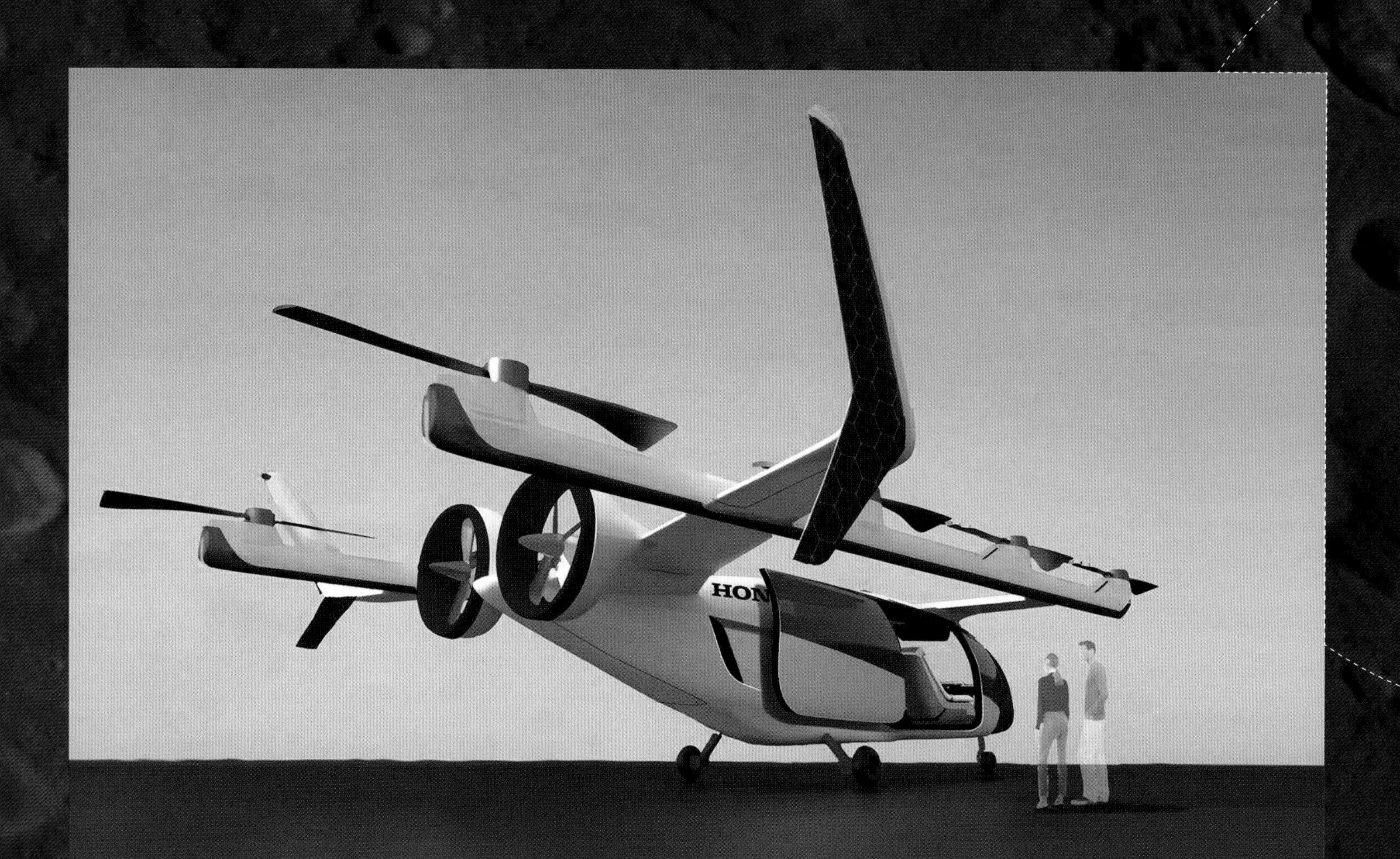

Honda is developing eVTOLS and other technology in preparation to assist NASA's moon ambitions. But whether Honda-made eVTOLS will actually fly humans around the moon remains to be seen.

The crater at center with the smaller craters within it is Clavius crater, where NASA discovered the dispersed presence of water.

Days after this and at the very end of July 2021, Blue Origin's protest of SpaceX's multibillion-dollar contract was rejected by the GAO, throwing Bezos's aims to a claim on the lunar lander project out the window, according to a statement from the U.S. watchdog organization. "NASA made [an] award to SpaceX for a total evaluated price of $2,941,394,557. After noting that SpaceX submitted the lowest-priced proposal with the highest rating, and that the offers submitted by Blue Origin and Dynetics were significantly higher in price, NASA also concluded that the agency lacked the necessary funding to make more than one award," read the GAO statement. According to the agency, NASA did nothing wrong and made no violations of regulations or laws when it decided SpaceX would become the sole recipient. "As a result, GAO denied the protest arguments that NASA acted improperly in making a single award to SpaceX."

After the ruling from GAO, Blue Origin expressed disappointment: "We stand firm in our belief that there were fundamental issues with NASA's decision, but the GAO wasn't able to address them due to their limited jurisdiction," said Bezos's aerospace firm, according to a *TechCrunch* report. "We'll continue to advocate for two immediate [lunar lander] providers as we believe it is the right solution."

Musk responded to the news in typical fashion, tweeting "GAO" followed by a flexed-arm emoji. A Blue Origin spokesperson declared that the firm would continue to pursue a way around the ruling, expressing optimism about the possibility of lawmakers writing a new provision in a Senate bill that would force NASA to choose two distinct companies to develop the HLS system.

Meanwhile, NASA's new spacesuits, called Exploration Extravehicular Mobility Units (xEMUs), which were also slated for completion in 2024, could no longer meet the deadline. The delays announced in NASA's August 2021 audit include the ISS Demo suit, two lunar flight suits, and an additional pair of qualification suits. "These delays—attributable to funding shortfalls, COVID-19 impacts, and technical challenges—have left no schedule margin for delivery of the two flight-ready xEMUs," read the audit. "Given the integration requirements, the suits would not be ready for flight until April 2025 at the earliest." NASA also said continued delays to the development of the Space Launch System and the HLS (which was paused during Blue Origin's protest) would " also preclude a 2024 lunar landing."

Three days later, on August 13, 2021, Blue Origin sued NASA in a major escalation of its struggle to get a piece of NASA's lunar lander program, according to an initial report from SpaceNews. The suit was filed in the Court of Federal Claims, which possesses jurisdiction over the bid protests that were initially examined by the GAO. Blue Origin's lawsuit called for a protective order to seal all the documents filed in its case, which was granted on August 16. The suit was sealed, Blue Origin claiming it contained "confidential, proprietary, and source selection information," along with "hearing transcripts in the bid protest."

However, in the request for sealed status, Bezos's firm referenced the HLS "Option A" award, the same lunar lander contract that NASA granted to SpaceX. "More specifically, this bid protest challenges NASA's unlawful and improper evaluation of proposals submitted under

the HLS Option A BAA," read the request while referencing the broad agency announcement of SpaceX's winning bid. "Blue Origin filed suit in the U.S. Court of Federal Claims in an attempt to remedy the flaws in the acquisition process found in NASA's Human Landing System," Bezos's firm told *SpaceNews*. "We firmly believe that the issues identified in this procurement and its outcomes must be addressed to restore fairness, create competition, and ensure a safe return to the moon for America."

With progress on SpaceX's contractual lunar lander development paused for ninety-five days during the GAO's initial adjudication, it seemed the additional delay to Elon Musk's Starship development put a pin in the entire Artemis program. As 2021 drew to a close, NASA's plans to launch Artemis I were officially delayed, along with every adjacent program, including the new spacesuits, the Lunar Gateway, and more. They would not move forward in January 2022, as NASA had speculated. "The Trump administration's target of [a] 2024 human landing was not grounded in technical feasibility," said NASA administrator Bill Nelson during a conference call on November 9, according to a CNBC report.

During the call, Nelson added that the Artemis I mission, which would be unmanned, would shoot for an orbital flight around the moon in spring 2022.

← NASA hinted that its perpetually delayed plans for the moon are linked to Jeff Bezos's continual lawsuits. Here he wears goggles owned by Amelia Earhart, which he carried into space.

→ In a surprise to the industry, NASA chose SpaceX as the sole recipient of its lunar lander development program, for which Musk's firm would continue to develop the Starship vehicle. Image: SpaceX

NASA

But he added that the first crewed mission on the SLS, dubbed Artemis II, would aim for May 2024. A follow-up mission, Artemis III, would return humans to the moon for the first time since Apollo in 2025 at the earliest.

At the end of September 2021, Musk vented about Blue Origin's continued interference in SpaceX's HLS Starship during a conference, stating, "You cannot sue your way to the moon," according to a report from *The Verge*. While he seemed to have a point, Amazon quickly released a list of litigations that SpaceX had filed against various federal and private entities during and after its rise to power as the corporate vehicle of the youngest space baron. Most notable among the defendants of SpaceX suits were the U.S. Air Force, for awarding launch service agreements to Blue Origin instead of SpaceX; Boeing and Lockheed Martin, in a dispute for providing satellite launch vehicles to the government; Northrop Grumman; and, of course, NASA, for selecting a company called Orbital Sciences Corporation for a lunar science mission before considering SpaceX's launch vehicles.

In all, the document consolidates twenty-nine of SpaceX's actions into three main categories: protests with the Government Accountability Office (GAO), litigation, and oppositions filed with the FCC. Notably, this list wasn't requested by the media. Amazon sent it to The Verge on its own initiative. Musk replied in a tweet to the news of Amazon's list, arguing that SpaceX used legal action to get into the figurative door of the private aerospace market,

↑ NASA's new spacesuits, called Exploration Extravehicular Mobility Units (xEMUs), were delayed indefinitely, putting all of Artemis in question.

→ Bezos argued that a wide spectrum of technology from a diverse collection of firms strengthened NASA's ambitions for the moon.

whereas Blue Origin was attempting to halt the market's progress. "SpaceX has sued to be *allowed* to compete, BO is suing to stop competition," read the tweet.

The list from Amazon was an exhaustive (albeit remarkably petty) counter to Musk's complaints. In a way, Bezos was telling his rival that while his lawsuits might be inconvenient for Musk and NASA, he (Bezos) had learned from the best. And, since some of SpaceX's actions happened in 2020, long after SpaceX was already a major player in the private spaceflight industry, Bezos wasn't entirely wrong. But, while the top two space barons bicker about who plays the game more unfairly than the other, space enthusiasts (and NASA itself), anxious for Artemis (and Starship) to return humans to the moon, have no clear picture of when the next human chapter of space travel will move out of the courtroom and back into the final frontier.

ARTEMIS

Blue Origin wanted NASA to include its version of a Human Landing System in the contract that was instead given solely to SpaceX.

THE RACE ITSELF

Cheers could be heard from a news team covering the test flight and landing of SpaceX's third test vehicle, called Starship SN10, as its engines shut down and the spacecraft maintained its upright position. But these cheers soon gave way to dread. During its powered descent, the potent Raptor engines had set the bottom portion of the vehicle aflame in what was suspected to be a dangerous methane leak. Commentators feared it might topple over as the structure of the vehicle weakened under the intensifying heat. Extinguishers fired at full blast, but the flames only grew larger.

For ten minutes, the world zoomed in on the growing inferno, waiting for the collapse and fall of the Starship, until suddenly everything disappeared in a bright glare, and then a massive orange plume filled the screen: The entire spacecraft had exploded in a colossal ball of fire, and the ever-flowing news cycle made a bizarre pivot to rewind. Major names across the space beat had already published articles congratulating Elon Musk on his company's first successful test landing of a next-gen reusable spacecraft that could return humans to the moon. Meanwhile, the reality was going

Behold the Mk1 rocket, shortly before Elon Musk unveiled it to the world. This event felt like the world was turning the page on the Space Race.

up in smoke in SpaceX's base at Boca Chica, Texas.

For a few minutes, the world existed in two incompatible realities.

But this wasn't wholly unexpected, which is why no one was anywhere near SN10 when it combusted into smithereens. And when the smoke finally cleared, all that was left under the partly cloudy sky on March 3, 2021, was the cone of what could have been the methane tank, as debris rained down on the smoldering landing area. This was the most dramatic ending of a Starship launch in 2021, but the procedure of try, fail, and fail better was already endemic to SpaceX's flight tests for Starship.

Elon Musk initially wanted Starship to be made of carbon fiber, but in January 2019, the CEO of SpaceX announced that he was pivoting to stainless steel. It's a heavier material, which means more fuel is needed to escape Earth's gravity. Musk also emphasized the thermal advantages of stainless steel, which would cut costs in a longer timeframe. But Starship has seen more design alterations since 2019. For example, the initial concept called for seven Raptor engines, which was later decreased to six.

Enter Starhopper

But before all these changes, SpaceX had to perfect the basic idea of a vehicle worthy of interplanetary travel that can "hop" from the ground into space and then land again. The Falcon 9 and Super Heavy rockets are powerful, but they weren't designed for long-term deep-space missions, where they'd need more fuel for contingent maneuvers and

↑ A destroyed Raptor engine from Starship SN10, whose explosion after successfully landing created two incompatible realities: one where it exploded and one where it survived.

→ The single Raptor-engine Starhopper tested key features of Musk's Starship, like reusability.

DESIGNATED
SMOKING
AREA
PLEASE

bountiful supplies, not to mention living space. These are among the reasons Musk introduced Starhopper, his firm's prototype rocket for the Starship.

Its maiden hop happened on July 25, a day after a glitch forced the ground team to abort an attempt slated for July 24. The abort was caused by excessively high chamber pressure on the Starhopper, reportedly the result of "colder than expected propellant," according to a Space.com report. The next day, and using a single Raptor engine, the Starhopper lifted above the ground a few minutes after midnight, reaching its primary objective altitude of 65 feet (20 m) and returning safely to Earth. "Starhopper test flight successful," Musk tweeted. "Water towers *can* fly haha!!"

The upgraded material was critical for this success. "Yeah, big advantage being made of high strength stainless steel: not bothered by a little heat!" added Musk in a follow-up tweet. Notably, this first untethered Starhopper launch came roughly six hours after the company's eighteenth robotic cargo launch to the International Space Station for NASA. This involved a Falcon 9 rocket lifting a Crew Dragon capsule to the orbital lab, the third flight for the capsule. The first stage of the Falcon 9 was also reused for the mission, which was its second flight.

Starhopper had made two tethered hops early in April 2019, which involved a slow, hovering maneuver; its second, and final, untethered hop happened on August 27, 2019. It rose to almost 500 feet (152 m), hovering a lateral distance of about 300 feet (91 m) after taking off at 6 p.m. EDT, and touched down on a separate landing pad after its fifty-seven-second flight.

Starhopper reached its flight altitude limit as mandated by the Federal Aviation Administration (FAA), the agency that grants licenses for test flights and launches. This final "hop" was far higher than the three earlier flights, and while the August flight was initially slated for August 26, this too was aborted just before liftoff, probably due to issues with the Raptor engine igniter, according to another Space.com report. But these few Starhopper flights effectively proved the Raptor engine's

"Water towers *can* fly haha!!"

—Elon Musk

← Elon Musk shared a video of the first untethered Starhopper test, with its single raptor engine.

→ Starhopper rocket is seen before SpaceX performs an untethered test of the company's Raptor engine. SpaceX's Raptor engines are the firm's answer to the Space Shuttle's.

↓ The Starship SN10 lifted off from SpaceX's Boca Chica base with high hopes for crucial data collection.

The Starship's design eschewed carbon fiber for a high-strength steel that could weather higher temperatures.

"Stainless steel is by far the best design decision we have made."

—Elon Musk

mettle, which opened the doors for the Starship and its first-stage rocket partner, the Super Heavy.

A Stainless-Steel Spaceship

With Starhopper tests completed, the baton of the second Space Race, as far as Elon Musk's aerospace firm was concerned, was officially handed off to the Starship prototypes. The first two, the Starship Mark 1 (Mk1) and Starship Mark 2 (Mk2), were built in separate facilities—Mk1 in Boca Chica and Mk2 on Florida's Space Coast—to foster internal competition at SpaceX. The first two Starships would be powered by three Raptor engines each, with test launches set for a rapidly approaching deadline. All eyes in the space industry turned to Musk as the next generation of spaceflight loomed large.

And on a starry Texas evening, September 28, 2019, Elon Musk revealed the first completed Starship vehicle to the world, only hours after his team of contractors, technicians, and engineers finished assembling the machine. The debut happened in an open-air shipyard near the Rio Grande River. It looked surreal behind the CEO, as he spoke to a group of a few hundred attendees, which included residents from Brownsville, Texas, and other nearby towns, in addition to SpaceX employees and the media. "This is the most inspiring thing that I have ever seen," noted Musk, wearing a black blazer over a T-shirt with jeans, about the tall, almost sentient presence of the massive next-gen rocket vehicle.

The crowd cheered as Mars felt like it had moved a little closer than ever before. We were finally making progress toward a future that we could touch. And it took the form of a giant, silver, 50-foot-tall (15 m) steel rocket. As the crowd took in the moment, Musk shared a handful of key features of the new vehicle. "Stainless steel is by far the best design decision we have made," he said during the debut event. Unlike carbon composites or aluminum-based materials, stainless steel doesn't become brittle. Additionally, at the unconscionably high temperatures of reentry, stainless steel doesn't melt until it surpasses 2,732°F (1,500°C). This means Starship only needs a comparably modest heat shield, composed of thermal tiles, to survive the threshold.

Beyond operational specifications, SpaceX's Starship also cuts costs by using steel. Carbon fiber costs up to $130,000 per ton, whereas stainless steel is only $2,500 per ton. "Steel is easy to weld and weather resistant," added Musk during the debut event. "The evidence being that we welded

← The Falcon 9 became the go-to rocket to lift SpaceX's Crew Dragon Capsule to the ISS.

→ SpaceX's partnership with NASA helped fund its Starship program.

this outdoors, without a factory. Honestly, I'm in love with steel."

But not everyone shared Musk's enthusiasm for steel and Starship. NASA had invested almost nothing in Musk's newest space vehicle but was heavily invested in crewed launches NASA contracted for his Falcon 9. So, the night before Musk's Texas Starship unveiling, then acting NASA administrator Jim Bridenstine made some sobering remarks that cooled the mood, expressing moderate concern about Musk's timeline. "NASA expects to see the same level of enthusiasm focused on the investments of the American taxpayer," said Bridenstine about SpaceX's implied overcommitment to Starship. "It's time to deliver." But Musk said his firm had invested only roughly 5 percent of its human resources to the production flow of Starship, with the lion's share of 6,000 employees hard at work on the Falcon 9 rocket, in addition to the Crew Dragon spacecraft—two critical elements of the commercial crew program.

Near the end of the Starship debut event, Musk answered questions from *Ars Technica* about his initial timeline for returning humans to the moon and his world-historical aims of taking a giant leap forward—to Mars. "It depends on whether development remains exponential," he said in an *Ars Technica* report. "If it remains exponential, it could be, like, two years," referring to a future moon landing.

While Musk was optimistic about sending Starship on a cargo trip to Mars by 2022 (when launch windows are ideal), he couldn't promise anything,

↑ NASA administrator Jim Bridenstine was ready for Musk to deliver on his promise to NASA.

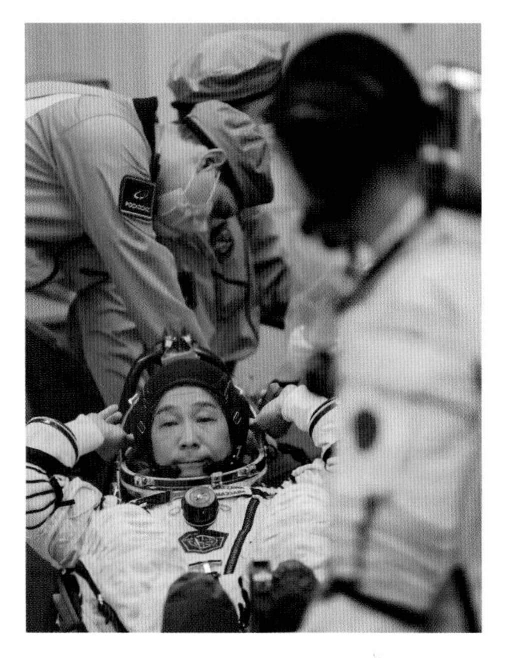

↑ Yusaku Maezawa has elected to become Musk's first deep-space flight customer.

↓ The Crew Dragon uses lithium hydroxide to "scrub" carbon dioxide from the air.

erring on the side of caution, should complications cause delays. And, compared to other, far less realistic timelines for reaching Mars, this conservative optimism was a rare thing. Much of Starship's funding came from satellite launches, with another $1 billion from private investors. But one investor of note, Japanese billionaire Yusaku Maezawa, planned to become Musk's first deep-space customer: hitching a ride to orbit the moon and then returning to Earth.

"I think we're able to see a path to getting the ship to orbit, and maybe even doing a loop around the moon," said Musk in the *Ars Technica* report. "Maybe we need to raise some more money to go to the moon or landing on Mars. But at least getting the Starship to an operational level in low Earth orbit, or around the moon, I feel like we're in good shape for that."

But once in orbit and on its way to deep space, the Starship will have to keep its crew alive. And, while designing the Crew Dragon spacecraft for NASA's crewed missions to the ISS, SpaceX built a thorough knowledge base for this.

"We definitely have learn[ed] a lot, and we would do it differently," said Musk during the debut event, contrasting his plans with previous lunar missions. "The Dragon life support system is not really all that renewable. It's basically mostly expendable." The Crew Dragon employs lithium hydroxide to "scrub" carbon dioxide that the crew exhales and produces by-products: lithium carbonate and water. These are sufficient for a four-person crew over a four-day journey, which means the

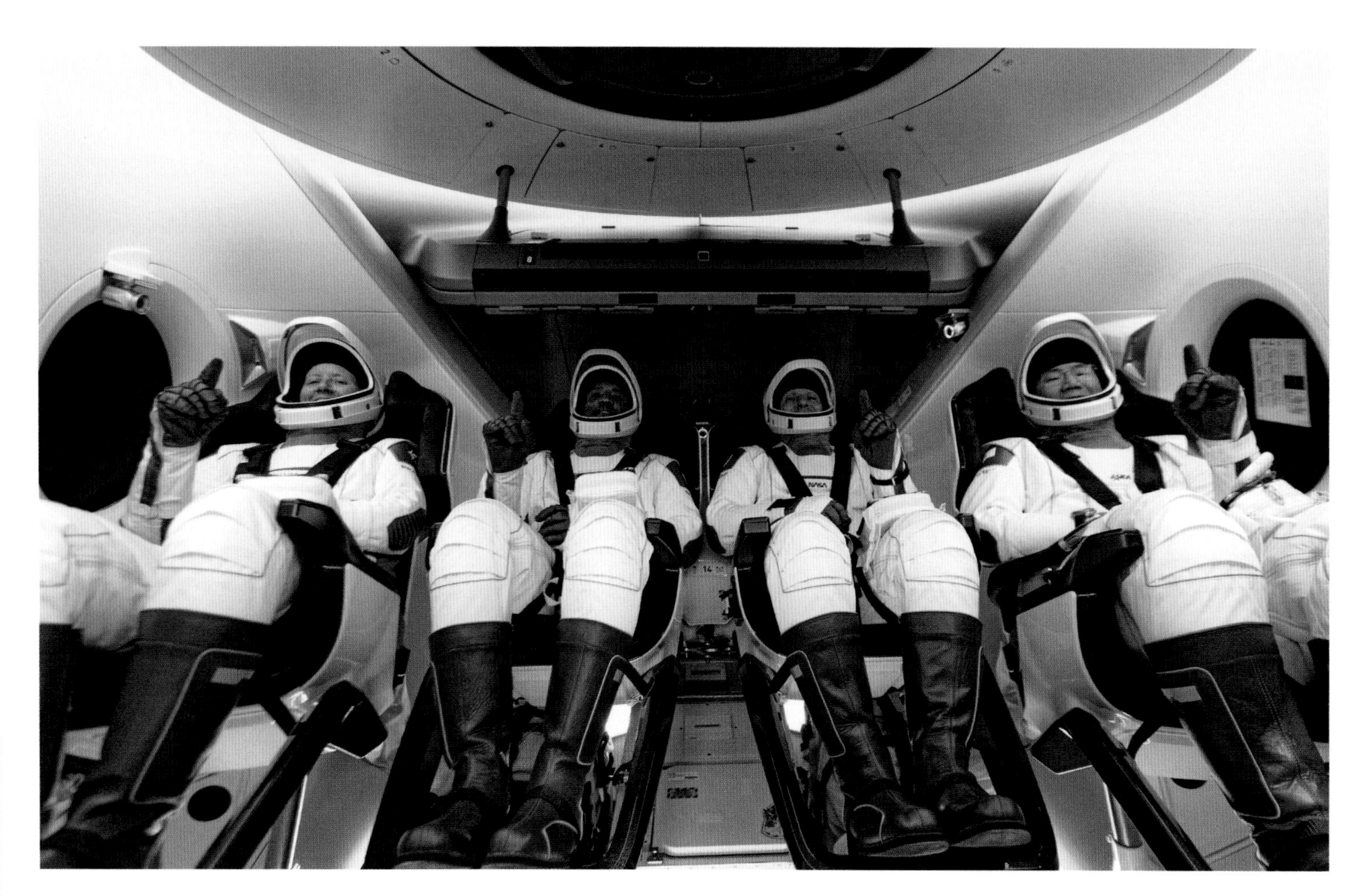

Sadly (for Musk), you can't get to Mars in a Tesla. Even this one, launched by SpaceX in February 2018, won't reach the Red Planet (because the trajectory was wrong).

process might support life for quick missions into lunar orbit and down to the surface of the moon. But getting to Mars is an entirely different proposition.

Answering Competition from NASA

The shortest trip to Mars is six months via conventional chemical propulsion (rockets), with a round-trip mission consuming up to two and a half years. In its finalized stage, SpaceX's Starship system will be capable of ferrying up to one hundred people to and from the Red Planet. A life support system of this scale would call for a regenerative life system that even Musk himself said will "take a bit of work" to realize. But while Starship had a long way to go before it could lift humans, it still seemed more likely to reach orbit before NASA's forthcoming Space Launch System (SLS), the agency's proprietary next-gen rocket for deep-space missions.

Earlier, in September 2019, Richard Shelby, a senior senator from Alabama, tweeted "Good News" about NASA's conjoining of five integral structures of the SLS's core stage. "This is the first time since the Apollo program that a rocket of this size has been joined together—a milestone accomplishment." This was an early step on the road to completing SLS, but NASA and its core-stage contractor, Boeing, would run into significant problems on the road to the initial test of the new launch system.

For starters, by 2014, NASA had invested tens of millions of dollars to modernize the Michoud Assembly Facility in Southern Louisiana, where the core stage was constructed. "They're throwing the money into this program, into places like Michoud, to make it very expensive to change course," said Peter Wilson, an adjunct international defense researcher at the nonprofit,

nonpartisan RAND Corporation, in a *Houston Chronicle* report. NASA has done this for other programs, like the James Webb Space Telescope, which launched in late 2021. And one reason might be to dissuade Congress from defunding programs every two to six years, when administrations and congresspeople are swapped in and out, along with pivoting policies on space program funding.

By 2019, NASA had spent more than $10 billion on the SLS launch system over a long development flow of five and a half years, and some have suggested that the lengthy development schedule reflected a lack of urgency among NASA officials. Flagship development programs like heavy-lift rockets designed to lift humans into deep space work well when funding is scant through the design phase, then receives a boost during development, and finally is reduced as flight production moves forward.

Breaking with this ideal scenario, Congress signed off on the SLS rocket program with a baseline investment of $2 billion annually and kept funding at this level plus inflation, more or less. While this is an incredible commitment to the job security of everyone involved in the project, it's not the most efficient way to develop an entirely new rocket system on a tight deadline. The SLS rocket core stage includes four Space Shuttle main engines and stands 212 feet (64.6 m) tall, with a 27.6-foot (8.4 m) diameter. In stark contrast, SpaceX's Starship prototype took a different approach.

The Starship Mk1 was 164 feet (50 m) tall, with a diameter close to 30 feet (9 m), which means the two vehicles were roughly the same size. However, neither were complete in this form—Starship is only the upper (or second) stage of its next-gen rocket vehicle, the Super Heavy; the SLS is accompanied by two parallel-mounted solid-rocket boosters, based on those used in the now-defunct Space Shuttle program. Most crucially, while the SLS core stage cost billions of dollars and took nearly a decade to complete, the Starship Mk1

← Artist concept of NASA's Space Launch System (SLS) solid rocket boosters firing their separation rockets and pushing away from the core stage. SLS, NASA's next-gen deep-space rocket for lunar missions, has suffered heavy delays and cost a large fortune.

→ NASA's SLS core stage test fire was one step in a long and arduous project that had already cost $10 billion by 2019.

The SLS rocket for Artemis I inside the Vehicle Assembly Building (VAB) at NASA's Kennedy Space Center in September 2021

Starship is only the second or upper stage of the fully stacked rocket. The bottom half, or first-stage rocket, is called the Super Heavy.

was built in a few months in 2019 and cost U.S. taxpayers almost nothing (not counting government subsidies and legal disputes between Musk and the agency, of course).

And the SN5 Starship prototype made its first hop in early 2020, less than a year after the vehicle was constructed. "If the schedule is long, it's wrong; if it's tight, it's right," said Musk during the Starship debut. And, as the flight tests of the prototype show, he's not wrong. But when a new engineering model enters the bottleneck phase of rapid testing, some shaky beginnings are to be expected. Months after the Starship debut, on November 20, 2019, SpaceX's Starship Mk1 prototype experienced a fatal error during preliminary testing at the firm's Boca Chica facility, blowing its lid mid-testing of its cryogenic system.

This was a standard test to confirm that the rocket could withstand the extremely cold temperatures of outer space. You don't want to be in a rocket undergoing explosive decompression before you have a chance to tweet about it (or at all, really), which is why it was actually good that SpaceX performed the test on the ground with no one aboard and within Earth atmosphere. The next step was Mk3 (Mk2 was discontinued when the Florida facility was deconstructed in 2020). Mk3 was also known as SN1 (serial number 1) and was destroyed during a pressurization test on February 28, 2020, at the firm's Texas base.

Video of the test from numerous vantage points showed the vehicle busting open at just about 11 p.m. EST in a sphere of high-pressure chaos from atop its test stand fueled by pure liquid nitrogen. No Raptor engines were installed, nor was there a nose cone, but the unscheduled rapid disassembly seemed to start near the bottom of the craft, which sent the top section into the air. When it landed, the resulting compression caused a second explosion. The next morning, little was left to see of SN1.

Naturally, there were no injuries from the explosion, and the CEO of SpaceX only tweeted commentary in the form of a video, with the caption "So . . . how was your night?" By day's end, word was that Musk's firm was already preparing the next iteration, Starship SN2, for test flights. This vehicle would come with enhanced welding techniques and production quality. But perhaps in anticipation of the two-for-two rate of test failures, Musk spoke about the challenges in designing an entirely new series of rocket vehicles. "Designing a rocket is trivial," he said during an onstage interview on February 28, 2020, just hours before the sudden end of SN1, according to a *SpaceNews* report.

Musk thought the real obstacle to rocket production lay in the steps immediately following design: namely, producing at volume and fully operational. "[T]he hard part is making it, and making lots of them, and launching frequently," he added in the report. A few weeks later, on March 8, 2020, Starship SN2 passed the cryogenic pressure test. "SN2 (with thrust puck) passed cryo pressure and engine load tests late last night," he tweeted the next day. At the time, Musk tweeted that SN3 and SN4 would see both "static fire & short flights" in reply to a Twitter follower who asked what was coming next after the successful cryo test. But neither would lift off—both had darker fates in store.

On April 2, SpaceX started tests of the SN3 vehicle, but the video footage

SpaceX's Boca Chica facilities
are a surreal sight to behold.

The SLS uses Space Shuttle engines. While adequate, they're also a forty-year-old design.

did not look promising: Deep into another pressure test at the Boca Chica launch site, the vehicle suddenly collapsed after what Musk claimed was an incorrect command input that dropped the internal pressure of the vehicle, causing the structure to collapse. "There are redundant pressure control valves," Musk tweeted about SN3's failure. "It's a new system and SN3 was simply commanded wrong. Rockets are hard. Good news is that this was a test configuration error, rather than a design or build mistake. Not enough pressure in the LOX tank ullage to maintain stability with a heavy load in the CH4 tank. This was done with N2," he added, referencing how this problem was already avoided with the SN2's successful pressure test.

On May 5, Starship SN4 fired its Raptor engine for the first time, raising hopes that it might be the first of the series to make a genuine test flight. Its single engine lit up the night sky during a brief static test fire. "Starship SN4 passed static fire," tweeted Musk after the successful test. The next test aimed for the SN4 to make an uncrewed test hop to roughly 500 feet (152 m). But these hopes were dashed when the vehicle exploded in a ball of hellish fire on May 29, at about 2:49 p.m. EDT. The test involved a short burn of its Raptor rocket engine in anticipation of the upcoming test hop.

But not all was lost. The day before SN4's end, SpaceX had officially received a launch license for Starship flights from the FAA. And, a month earlier, NASA had chosen SpaceX's Starship as one of three primary commercial vehicles to eventually lift astronauts to the moon as soon as the agency's Artemis program finally got off the ground. But most crucially, the production line for Starships was starting to gain speed. After a series of new hires at the firm, Musk expressed his goals to develop the vehicles at an insane rate: at least one completely new Starship *per week*.

Per week!

By contrast, NASA's Space Launch System had been in development for a decade, during which its contracted private builder, Boeing, had only completed its single core stage. Additionally, once NASA launches the SLS, each core stage will be abandoned in the ocean after just one use. And, instead of designing new engines for a new century of space travel, NASA is using the forty-year-old main engine design from the Space Shuttle program. As of early 2022, the completed SLS still hadn't lifted off. By contrast, Elon Musk's plans to build one or two Starships every week in 2020, at just $5 million each, seemed unspeakably overzealous.

"That's fucking insane", said Eric Berger in an interview with Elon Musk at *Ars Technica*. "Yeah, it's insane," replied Musk. "I mean, it really is," repeated Berger. "Yeah, it's nuts," affirmed Musk.

"I think we need, probably, on the order of 1,000 ships, and each of those ships would have more payload than the Saturn V—and be reusable."

—Elon Musk

The way to space begins where the road ends, at SpaceX's Starship launch.

This was a real conversation. "The conventional space paradigms do not apply to what we're doing here. We're trying to build a massive fleet to make Mars habitable, to make life multi-planetary. I think we need, probably, on the order of 1,000 ships, and each of those ships would have more payload than the Saturn V—and be reusable." Musk went on to emphasize the scale of interplanetary transport needed to seed a settlement on Mars.

And Musk estimated that the order of magnitude to make a truly self-sustaining community on Mars is "probably not less than a million tons." But to make this happen, Musk needed more than a fresh batch of new hires. To "make something at reasonable volume, you have to build the machine that makes the machine, which mathematically is going to be vastly more complicated than the machine itself," said Musk in the *Ars Technica* report. And the work of building this machine would be absolutely grueling.

Twelve-hour workdays, with four-day weekends, were allotted to SpaceX engineers. After this, the work schedule was twelve-hour shifts and three-day weekends. This meant Boca Chica could function at full capacity twenty-four hours per day, every day of the week. While small solace to workers building on the razor's edge of burnout, SpaceX did include hot meals for everyone every three hours. So, there's that.

But by the time Starship SN5 was constructed, SpaceX had completed its first orbital launchpad, in June 2020. The latest prototype vehicle was moved out for tests without a nose cone on July 1, successfully passing its cryogenic proof test in the evening, signaling a quick recovery for the growing aerospace firm that had only one month earlier

SPEED LIMIT
10
STOP
PENSKE
Truck Rental
SPACEX

The key to making humanity a multi-planetary species is to build Starships at scale. Musk reportedly fell "in love" with 304L stainless steel.

Starship SN8 landed exactly where it was supposed to—just way too fast.

suffered an explosive engineering error when SN4 burst into glorious fire.

A cryogenic proof involves filling the Starship's propellant tanks with liquid nitrogen and then pressurizing them to full flight levels. Once these flight pressures are reached, hydraulic pistons (also called a thrust simulator) push against the bottom of the vehicle to create the forces expected from a real Raptor engine fire.

A few months later, SN5 made its first test hop on August 4, 2020. Firing its single Raptor engine, it soared close to 500 feet (152 m) before descending for a soft touchdown with zero explosions. The mood was ebullient as Musk's firm seemed to be making rapid progress in a year rife with suffering, eight months deep into the COVID-19 crisis.

Only one month later, Starship SN6 performed the very same hop, rising to essentially the same altitude on September 3, 2020, and then touching down without a hitch. And while the goal for the Starship prototype series remained a daunting altitude of twelve miles (19.5 km), the firm had yet to launch the Starship vehicles beyond the 500-foot (152.5 m) altitude. One might expect SN7 to be the ideal successor for a higher-altitude flight, but it no longer existed. Why? Musk had something else in mind: its intentional destruction. On June 23, 2020, SpaceX took the massive tank of SN7 beyond its limits by filling its tank with liquid nitrogen until it surpassed its maximum containment pressure, coughing billowing clouds of white nitrogen into the atmosphere.

It's worth noting that these events were and continue to be covered in exquisite detail by a team of journalists on the ground at Boca Chica from NASASpaceflight.com, who shared indispensable video footage of every test fire and test flight of SpaceX, in real time. Without their commitment, the world would not have early access to videos and data from the events and their oft-explosive endings. When SN7 ruptured, for example, a video captured by "Mary" (also called BocaChicaGal) revealed how the tank popped and collapsed, subsequently falling into expanding plumes of nitrogen.

Musk said that SN7 leaked, and did not explode, possibly thanks to the firm's pivoting in construction materials from 301 stainless steel to 304L, according to a Space.com report. "Tank didn't burst, but leaked at 7.6 bar," tweeted Musk. "This is a good result & supports [the] idea of 304L stainless [steel] being better than 301. We're developing our own alloy to take this even further. Leak before burst is highly desirable."

Starship Test Flights

On December 9, 2020, the time had finally come. At 5:45 p.m. EST, SpaceX's SN8 made the prototype's first high-altitude test flight from the facility near Boca Chica, Texas. The aim of the flight was to fly at least 7.8 miles (12.5 km) high, execute a confection of complicated aerial maneuvers, and then land safely on the launchpad. One of the impressive airborne stunts included a "bellyflop," mimicking the motion required for a fully operational Starship during reentry. At 165 feet (50.5 m) in height, the SN8 successfully nailed every one of these goals. All except the last.

Six minutes and forty-two seconds following liftoff, Starship SN8 landed exactly where it was supposed to, but at an excessive, unrecoverable velocity, and exploded. But this didn't sour spirits for Musk, who celebrated the first

← SpaceX launches Starship SN9 for a test flight from its facilities in Boca Chica on February 2, 2021.

→ SN9 explodes into a fireball after its high-altitude test flight. The incident offered SpaceX engineers a bounty of data to help enhance its next prototype launch.

high-altitude flight by taking to Twitter. "Fuel header tank pressure was low during landing burn, causing touchdown velocity to be high & RUD, but we got all the data we needed! Congrats SpaceX team hell yeah!" he tweeted.

"Mars, here we come!!" added Musk in a follow-up tweet. The explosive failure was more likely than not: Musk had estimated a two-in-three chance that SN8 would return not in one piece, but several. This is because the flight was the most demanding and complex Starship prototype test ever. And SN8 has a full nose cone, unlike its predecessors. They appeared far less impressive by comparison, like stubby silos. SN8 was powered with three Raptors and employed body flaps to stabilize its attitude during flight.

It's difficult to overstate the significance of SN8's initial flight: This was the first time Starship proved it could not only launch high into the sky but also guide itself down to the correct landing region (although at excessive velocities). "SN8 did great!" tweeted Musk of the productive test. "Even reaching apogee would've been great, so controlling all [the] way to putting the crater in the right spot was epic." And, as expected, SpaceX was already preparing to launch SN9, the next Starship prototype.

On February 2, 2021, at 2:25 p.m. EST, Starship SN9 lifted off from SpaceX's Boca Chica base and flew more than six miles high (10 km). Like its predecessor, it performed the signature bellyflop and began its descent. Predictably, it didn't slow its velocity enough for a soft landing, in what was (perhaps euphemistically) called a "hard" landing. But this wasn't, strictly speaking, a failure.

While SN9 did explode upon impact, it still offered SpaceX engineers a bounty of data to help the firm enhance its next prototype launch. It was initially supposed to launch the previous week, in late January 2021, but the firm was forced to wait for the FAA's clearance. Needless to say, Musk was not pleased. "Unlike its aircraft division, which is fine, the FAA space division has a fundamentally broken regulatory structure," tweeted Musk. "Under those rules, humanity will never get to Mars."

But, perhaps unknown to Musk, the FAA's decision-making process was already undergoing rapid evolution, according to a tweet from *SpaceNews*. And the agency granted Musk approval only days after its previously scheduled launch, saying in a February 1 statement that SpaceX "complies with all safety and related federal regulations and is authorized to conduct Starship SN9 operations in accordance with its launch license," according to NASASpaceFlight.com.

Then came one of the most bizarre and exciting events for SpaceX's series of Starship prototypes: the test flight of SN10. It was 6:14 p.m. EST on March 3, 2021, when the prototype lifted off after initial delays. As it made its descent post-flip, it became obvious that the Starship was actually slowing down enough to appear less like an impending

crash than a viable landing. The only concern was the extra flames lapping the bottom edge of the colossal silver spacecraft, which didn't relent as the craft was enveloped in dust and exhaust upon touching down on the Boca Chica base's landing pad.

As the dust cleared, all seemed very nearly well, except that the fires hadn't extinguished. It appeared only a matter of time until it tipped over; after ten minutes, anyone still watching the livestream saw it explode gloriously—like post-credits bonus content, except this was really happening. It was a bizarre situation for the media: In a world of nearly instant posting, nearly every outlet had already published stories declaring the historic landing a complete success and a major step on the road to the moon, Mars, and beyond.

As senior editor at Interesting Engineering, I appeared to be one of the first to notice, immediately updating my live coverage to reflect the new surprise ending, which led to a Google search revealing that the SN10 had both landed successfully and exploded a few minutes later. Like some quantum experiment, for several minutes, the world was divided into two incompatible realities. Of course, all news coverage eventually backtracked and corrected their coverage to correspond to the new reality, but the event possibly foreshadowed something strange about the future of spaceflight: The proliferation of incompatible realities as a function of access resembles how the narrative of reality might diverge between, say, astronauts living on Mars and our interpretation of it on Earth.

Rekindling the Space Race

Time will tell if this is how things unfold, but SpaceX was clearly continuing

↑ Concerns deepened in the Boca Chica area following SN11 surrounding the potential environmental toll of constant Starship launches. Local residents were warned not to tamper with any raining debris from failed Starship launches.

undaunted, having previously launched its first human crew of astronauts aboard the Crew Dragon capsule atop a Falcon 9 rocket to the International Space Station nearly a year earlier on May 30, 2020. The Demo-2 mission went forward after an earlier weather delay, lifting NASA astronauts Bob Behnken and Doug Hurley to the ISS and docking with the station at 10:16 a.m. EDT, after which the pressure was equalized for three hours before the astronauts could transfer from the Crew Dragon capsule to the ISS itself at 1:22 p.m. EDT.

This marked a major and final step for SpaceX to receive certification from NASA for human spaceflight, but for Musk, it wasn't enough. Only landing on the moon and Mars would serve as a viable starting place, which is why Starship prototype tests went on. Following several delays, Starship SN11 soared into the Texas skies on the morning of March 30, 2021, at 9 a.m. EDT. It rose to an altitude of 6.2 miles (10 km) and then began its descent through intense fog. Surprisingly, six minutes post-liftoff, the broadcast cameras from SpaceX cut their feed of the event: "Looks like we've had another exciting test of Starship Number 11," said the firm's launch commentator John Insprucker, according to a Space.com report.

"Starship 11 is not coming back; do not wait for the landing," added Insprucker ominously. The big secret? SN11 had exploded during descent, scattering debris throughout the surrounding area, a mishap that aroused the concern of environmentalists worried SpaceX's mega-rockets might pose a danger with "pollutants" from

↑ The two NASA astronauts aboard the Crew Dragon Capsule before the Demo-2 mission

↓ NASA selected SpaceX's Starship to be its Human Landing System for the forthcoming Artemis missions to the moon.

It was beginning to look like Musk actually had a shot at completing his new Mars rocket.

exhaust or debris. The ruins of the exploded space vehicle had littered an area spanning an entire kilometer (3281 feet), penetrating into state and federally managed land, according to a document from the U.S. Fish and Wildlife Service (USFWS).

"SpaceX worked closely with USFWS and other agencies in retrieval of explosion debris to minimize impacts to wildlife and sensitive habitats," wrote the agency in a report from *The Rider*. SpaceX took the lead in recovering all remaining debris, centering on a collaborative strategy to minimize long-term environmental damage. Giant metal scraps from SN11 were lodged in the ground and the tidal flats near a nearby highway. Tractors were witnessed pulling the big chunks out, according to the report.

But even the cleanup process was seen to perturb the environment. "Access in the way of constant ingress/egress [that is, entrance and departure of the area] to retrieve debris materials created disturbance and unintentional paths across the landscape," wrote Aubry Buzek, a public affairs specialist for the USFWS, in an email to *The Rider*. In response to ongoing outcry from the public and other concerned citizens, SpaceX opened a hotline for everyone to call and report any portentous debris, advising the public not to handle any of the Starship remains scattered across the countryside.

This led many to envision a future for spaceflight that looked less utopic and glorious than grim and undesirable. While NASA announced a $2.9 billion contract to leverage a variant of SpaceX's Starship to be the crewed Human Landing System (HLS) for forthcoming Artemis missions to the moon only two weeks after SN11's calamitous end, local

↑ SpaceX's Crew-1 mission upped the ante, sending four astronauts to the ISS. From left: mission specialist Shannon Walker, pilot Victor Glover, and Crew Dragon commander Michael Hopkins, all NASA astronauts, and mission specialist Soichi Noguchi, Japan Aerospace Exploration Agency (JAXA) astronaut.

→ The Crew-1 astronauts launched in the dead of night.

← The Crew-2 astronauts found that the media's interest in astronaut interviews had been fully rejuvenated.

→ Rolling out the Crew-2, attached to the Falcon 9 rocket

residents and onlookers feared more serious incidents could happen when SpaceX began launching fully stacked Starships, propped up by the Super Heavy rockets.

But the testing continued. Starships SN12, SN13, and SN14 weren't yet completed mid-2021, so SpaceX skipped to SN15, which featured improved software, engines, and vehicle structure. On May 5, 2021, at 6:24 p.m. EDT, the SN15 prototype lifted off after getting an FCC permit to operate a Starlink dish on the vehicle. After reaching its target altitude, the Starship began its descent, performing the bellyflop maneuver with perfect execution. Amid the bated breath of all viewers, it landed softly, with zero explosions. This had never happened before. And, while there was still close to a three-story blazing fire near the bottom of the vehicle after landing, it seemed completely extinguished within an hour.

This marked the first-ever landing of Elon Musk's new Starship prototype after a full test flight. Without a doubt, the second Space Race felt a little bit

NASA

hotter. But soon after Starship SN16 was rolled out to the Boca Chica facilities' "rocket garden," Starship testing was halted amid legal disputes from Jeff Bezos's Blue Origin against NASA, according to a *CNBC* report on April 30. Sadly, Starship testing remained on hold until the disputes between NASA, Blue Origin, and SpaceX came to an end on November 4, 2021. The resolution was a colossal win for SpaceX, but it had forced NASA to officially delay its Artemis timeline.

SpaceX's Crewed Mission Launches

But not all was lost. While SpaceX couldn't move forward with Starship test flights, it could move forward with operations related to the Falcon 9, Crew Dragon, Starlink satellites, and more. In fact, what's perhaps most remarkable about SpaceX's operational architecture is that its Starship advances were funded by ongoing Falcon 9 launches. On November 15, 2020, the firm had launched its second crewed mission, called Crew-1, at 7:26 p.m. EST from Pad 39A at NASA's Kennedy Space Center.

The crew included three NASA astronauts and one astronaut from Japan's space agency, JAXA.

Before the hour was up, SpaceX's Crew Dragon had accelerated to 16,777 mph (27,000 km/h) and separated from the first-stage Falcon 9 rocket on its way to the ISS. The following year, on April 23, 2021, SpaceX launched its third crewed and second fully operational mission to the ISS at 5:49 a.m. EDT, this time ferrying four astronauts: two from NASA, one from JAXA, and the first-

↑ The Crew-3 astronauts, posing in Florida, pre-launch

→ SpaceX's first all-civilian mission, Inspiration4, had four non-NASA astronauts. From left: Jared Isaacman, Christopher Sembroski, Hayley Arceneaux, and Dr. Sian Proctor.

ever astronaut from the European Space Agency (ESA) to fly with SpaceX. There was also a post-launch news conference. Elon Musk (wearing a bandana) said he had high hopes to transform humanity into a "spacefaring civilization and a multi-planetary species," according to an Interesting Engineering report.

The astronauts of Crew-2 checked in from orbit, two hours deep into the twenty-three-hour trip to the space station, according to a tweet from NASA that displayed an awe-inspiring view. And, it would soon be succeeded by SpaceX's fifth overall crewed launch, the Crew-3 mission, which was also the fourth flight for SpaceX's Commercial Crew Program in partnership with NASA, and lifted the 600th person into space in sixty years of human spaceflight. The Falcon 9 rocket launched at 9:03 p.m. EST, on November 11, 2021, from NASA's Pad 39A at Kennedy Space Center. While the mission was initially delayed because of unfriendly weather, it was also pushed back when one of the Crew-3 astronauts sustained an unspecified injury before the launch.

This fifth crewed launch happened only two days after the Crew-2 astronauts returned to the planet, splashing down after a successful stay on the ISS on November 9, at 10:33 p.m. EST. Notably, on their way back, the Crew-2 astronauts enjoyed unbeatable views of the aurora borealis. But something had changed this time around. Between the Crew-2 and Crew-3 missions—the second and third

The symbolic significance of "Captain Kirk" going to space was for many the highlight of the year. After touchdown, Shatner spoke about how thin the atmosphere as a "thin blue veil" is between life on Earth and death in the black abyssal depths of space.

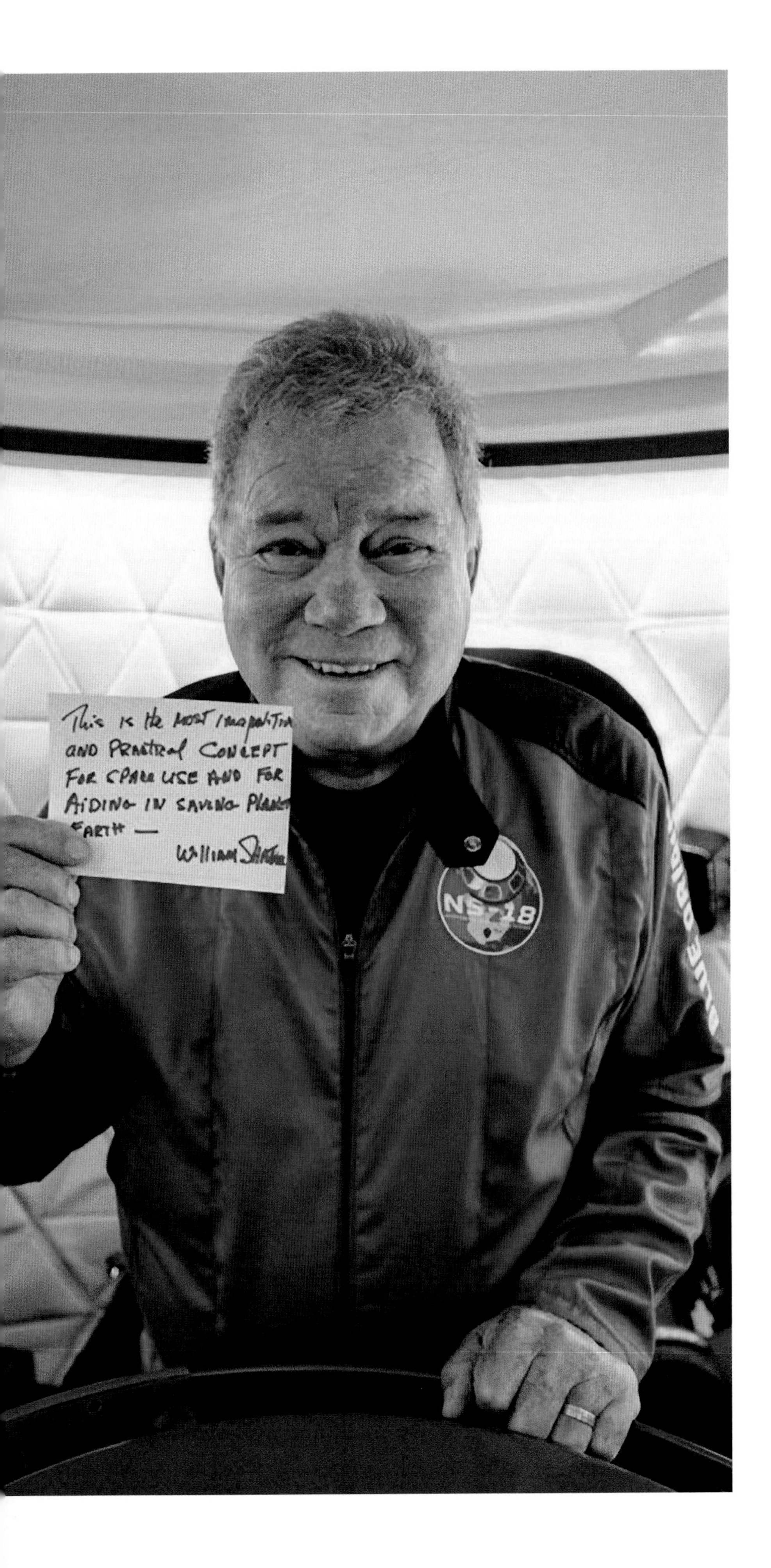

flights of the Commercial Crew Program with NASA, but SpaceX's third and fifth crewed flights since its initial Demo-2 mission—there was a fourth launch that broke with convention of space travel. Up to this point, it had become expected for space missions to be crewed by fully trained astronauts. But the very enterprise of space travel had just undergone an executive transformation that would forever change the purpose of space travel.

William Shatner Finally Enters the Final Frontier

On September 15, 2021, at 8:02 p.m. EDT, SpaceX made its fourth overall flight to space, but without any professional astronauts aboard. The first all-civilian mission (called Inspiration4) saw a Crew Dragon lift four occupants, including data engineer Chris Sembroski, physician assistant Hayley Arceneaux, geoscientist and science communication specialist Sian Procter, and the billionaire Jared Isaacman, who chartered the private spaceflight with Elon Musk. Space tourism had officially begun: A billionaire had forked over the cash for another billionaire's private aerospace firm to lift him and a few lucky others beyond the atmosphere of Earth.

This all-civilian mission began under a frame of charity, when a SpaceX press release revealed that the billionaire Isaacman, who is also a trained pilot and chief executive officer of Shift4 Payments, committed $100 million to St. Jude Children's Research Hospital, with aims of raising another $200 million by inviting civilian astronauts to give $10 donations to St. Jude. And this donation became the entry fee for the all-civilian mission. And that gets at a core feature of space tourism: a rapidly growing industry where billionaires voluntarily pay

← It's easy to see the aesthetic appeal of OAC's Voyager Station. But it might not be the "blank slate" for humanity that many would prefer.

↓ Top-paying space tourists will even have private access to a "space walk" terrace.

their way into space, while the rest of us enter what's effectively a space lottery.

Less than a month after Inspiration4, Blue Origin did something of much greater symbolic relevance to signal the dawn of a new age of private space travel. For nearly half a century, when people thought of starships exploring deep space, they thought of one ship, and one man specifically. His name is Captain Kirk, and in fact, no one has ever met him, because he isn't real. But the renowned actor behind the character, William Shatner, agreed to become the first *Star Trek* actor to actually become an astronaut.

On October 13, 2021, Blue Origin's New Shepard rocket lifted Shatner and three other civilians to the very edge of space around 10:50 a.m. EDT, according to a live webcast published on Blue Origin's YouTube channel. At ninety years old, Shatner became the oldest person to take a ride in a rocket into outer space. Which, among other things, means he made the sci-fi franchise's core message come true: He boldly went where few had gone before.

The New Shepard rocket was NS-18, and Jeff Bezos personally escorted all four crew members to their capsule. While the launch was at first slated for liftoff at 9:30 a.m. EDT, it was delayed and then placed on hold at 10:18 a.m. EDT. Thirty minutes later, at 10:50 a.m., the rocket finally began its short journey to the edge of space. Roughly two and a half minutes into the historic flight, zero-gravity conditions began to kick in, which triggered cheers from all aboard, including Shatner.

The crew capsule then separated from the booster three minutes post-liftoff. Thirty seconds later, the crew of this now world-historic flight officially entered space, rising to a maximum altitude of 350,000 feet (106,680 m) before its acceleration turned negative, signaling the beginning of a slow return to Earth's surface. The booster rocket that lifted Shatner to space has a cylindrical and more aerodynamic body, which enabled it to fall much more rapidly through the planet's atmosphere, touching down with zero explosions approximately seven minutes

OAC's Voyager Station could become a major commercial hub and jumping-off point for interplanetary commerce.

and twenty seconds after liftoff, ending the fourth jump to space for the New Shepard spacecraft.

William Shatner was among the ninety-sixth through ninety-ninth astronauts in space. After his short journey, he had some big feelings to convey. "This is like nothing I've ever experienced before," said Shatner, during Blue Origin's webcast. A small yet incredibly enthusiastic crowd had gathered to welcome him, including Bezos himself, who towered over the deeply moved actor as the latter attempted to explain that the thin blue line between life on Earth and death in space is so unspeakably thin and fragile that it's practically a fictive invention of society.

This flight into the black abyssal depths of space wasn't quite the jump to warp speed most of us have come to expect from the captain of the USS Enterprise, the fictional ship in *Star Trek*, but the giddy feeling post-landing was absolutely palpable. "It's time Captain Kirk actually physically got up into space," said Mayor Becky Brewster of Van Horn, a rural town of just about 1,800 people near Blue Origin's facilities in West Texas.

Arguably, the rise of space tourism transforms space into a physical analog of wealth disparity: The few richest people in the world can opt in to space tourism and enjoy something the rest of humanity could never touch, almost exclusively because of the conditions of our birth. Of course, it fits nicely into decaying mythology about boot-strapping success in America, where anyone willing to work hard can lift themselves out of poverty. Now, with space tourism, a new American mythology is born: You can lift yourself all the way to space—if you work hard. And even if you can't, so long as you're trying, it makes the few who can afford to buy a ticket to space look far more aspirational.

Artificial Gravity

This reality is most apparent with the growing circle of start-up companies unfolding in the budding space tourism industry. Perhaps most notable among them is the Orbital Assembly Corporation (OAC), which on January 30, 2021, announced its goal of launching the most ambitious project in space tourism: a large-scale, habitable "space hotel" in low-Earth orbit called the Voyager Space Station (VSS). The design is reminiscent of the von Braun concept for space stations, where a wheel-shaped (or doughnut-shaped) structure houses a habitable environment and spins at an angular velocity high enough to generate artificial gravity for everyone inside. At face value, this is a great idea because some semblance of gravity is essential for maintaining health on long-term space missions.

The closest cultural analog is the fictional space station depicted in the 1968 film *2001: A Space Odyssey* by Stanley Kubrick. But, instead of a silver screen prediction of what life might have been like twenty years ago, OAC is creating something in real life: a ring-shaped station with a diameter of 650 feet (198 m) supporting a level of gravity comparable to that of the moon. Revealingly, while the firm wasn't against the idea of including scientific and other strategic (or military) uses for the station, OAC envisions Voyager as a commercial hub. In other words, instead of a blank slate upon which humans could build a fundamentally novel form of democratic, space-based society, OAC wants to create a private platform specifically for the *business* of space travel to flourish,

If we're going to live and work in space, we'll need to create artificial gravity, and circular structures are a way to do this via centripetal force.

according to the company CEO and president John Blincow, who spoke during a webcast on YouTube.

"Our vision is to create a space construction company for the design, manufacture, and assembly of large structures in space, including commercial space stations, space solar platforms, and propellant depots," read the company's introduction in a press release. "To achieve this objective, we developed several design patents for in-space assembly robots."

As was the case with Musk's Starship production line, there's simply too much work to be done in scaling a space economy to bring an entire population of highly qualified engineers to bear. Instead, the core innovation is developing the machines that make the machines. "To enable a robust, human-centered space economy, our capabilities are geared toward the eventual construction of a Voyager Space Station (VSS)," said OAC in its initial offering statement, published on the U.S. Securities and Exchange Commission's website. "We plan to build the rotating space station in stages, starting with a small-scale station demonstration, and one or more free-flying microgravity facilities, utilizing VSS components."

"This will be the next industrial revolution," said Blincow, in the live webcast of OAC's event. For a long time, the biggest impediment to building industry in space was the cost. It was "about $8,000 per kilogram [roughly $17,600 per pound] for a long time," said Tim Alatorre, the COO, CFO, and VP of Business Administration and Habitation Architectural Design of OAC. "But with the Falcon 9, you can do it for less than $2,000. And as Starship comes online, it will only cost a few hundred dollars" per kilogram.

To build the station, OAC will use a Structure Truss Assembly Robot (STAR), which will operate as "the structural backbone of future projects in space," said Tim Clements, OAC's fabrication manager. The prototype assembly robot will construct a truss section that's approximately 300 feet (91 m) long in just ninety minutes, added Clements. Called DSTAR, the robot "weighs almost eight tons in mass—consisting of steel, electrical, and mechanical components." And, as an incredible bonus, some of us might get front-row seats to the construction process.

An observer drone will monitor the construction process in real time, offering the OAC team "eyes on the job site," explained Alatorre. "The observer drone . . . can perch on existing craft, [and] can also be fully reusable and can fly and have a free-flight mode on extended missions." Most crucially, the firm will

↑ Artist concept of the Voyager probe before a nebula in deep space

← It's pretty clear that the aim of Voyager Station is to serve paying customers with maximum comfort, in space.

affix a virtual reality (VR) headset on the robot, allowing remote control. And it could be critical for such a meticulous job since the main feature of artificial gravity has never been successfully implemented in space before.

"The gravity ring is going to be a key technology demonstration project that we plan to build, assemble, and operate in low-Earth orbit in just a few years' time," said Jeff Greenblatt, OAC's cofounder and VP of Science and Research, who is also a former staff member at NASA's Ames Research Center. "The company also plans to use an orbital version of the DSTAR called the PSTAR, which stands for Prototype Structural Truss Assembly Robot." Once the final Voyager Space Station is complete, it will hold up to 400 people, and the 650-foot (198 m) gravity ring will be internally portioned into twenty-four integrated habitation modules, every one of which will be roughly 65 feet (19 m) long and 40 feet (12 m) in diameter.

Of course, artificial gravity allows for operational toilets and showers and even running, jumping, and other relatively simple forms of physical exertion (although it will probably feel unnatural at first). The Voyager station will also have electrical power, air, and water, along with a kitchen module, and even a gym for sports and special events. Notably, the government and private firms will be able to rent the facilities to train astronauts for lunar industries. "It will be a springboard for entrepreneurs to plan tourist activities in space," said Greenblatt.

"We don't want the Voyager experience to be like being in an attack submarine in combat—so we're [building] for comfort," said Tom Spiker, the cofounder and CTO of OAC, who also worked on the legendary deep-

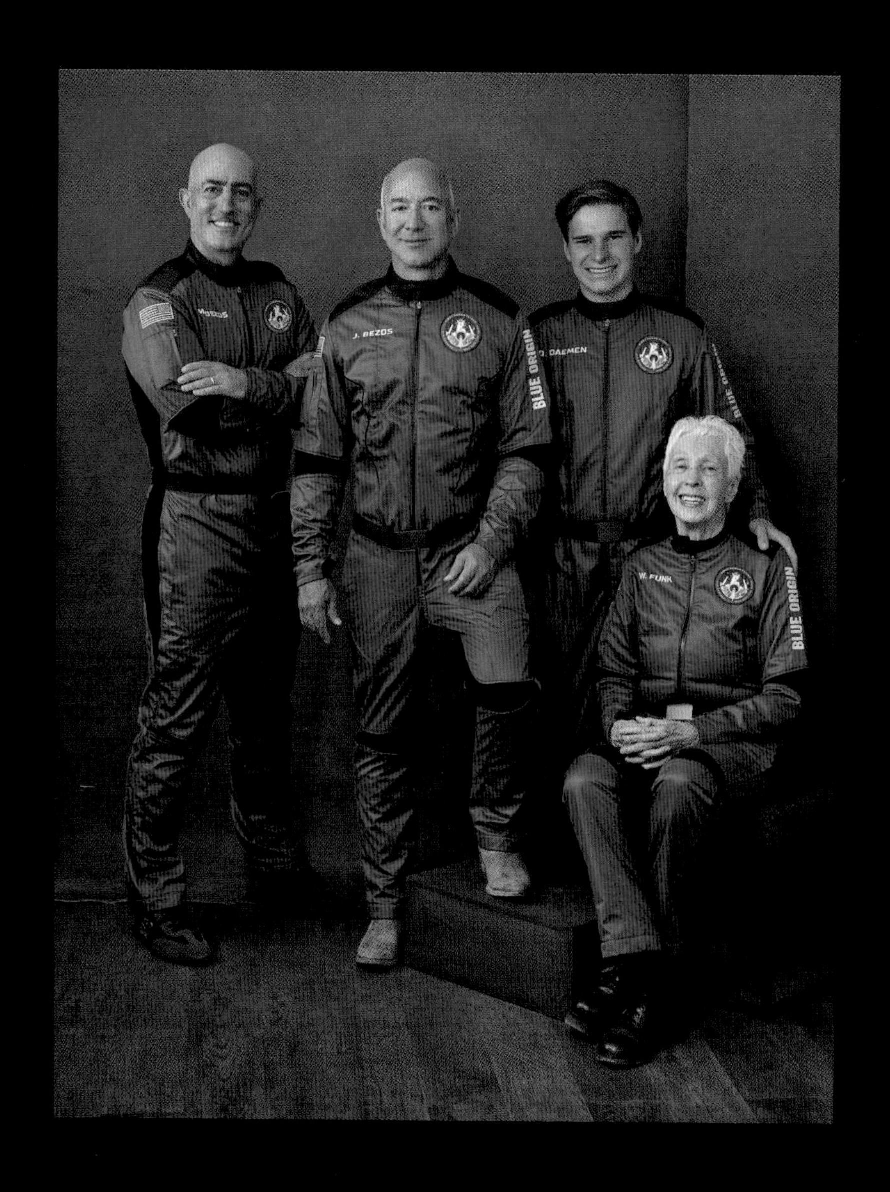

On July 20, 2021, Blue Origin's New Shepard rocket launched Jeff Bezos; his brother, Mark Bezos; 82-year-old aviation pioneer Wally Funk; and 18-year-old Oliver Daemen into space.

space Voyager probes, Cassini, and more while at NASA's Jet Propulsion Laboratory (JPL). "It's a bit smaller than the length of the U.S. Capitol Building" and spins at 1.25 rotations per minute. Most unspeakably luxurious on the growing list of space amenities is the option some visitors will have for a spacewalk, courtesy of their own private airlock, "where the only thing between you and the universe is a faceplate."

Space Tourism Holds a Lottery

The Orbital Assembly Corporation has already assembled a collective of investors to fund the gravity ring, Voyager Space Station, and more. But it's important to note that the primary goal of this station, and space tourism broadly, is less science than "tech," which means it's hard to say who among casual space enthusiasts the industry will ultimately serve. For now, it seems to serve the exceptionally well-off. Jeff Bezos launched himself to space in his New Shepard rocket on June 20, 2021, at 9:14 a.m. EDT, and the launch vehicle's crew capsule was named "RSS First Step."

He was joined by his brother, Mark Bezos, in addition to an eighty-two-year-old aviation pioneer and an eighteen-year-old student named Oliver Daemen, whose father paid for his ticket, making him Blue Origin's first paying customer.

But a more conventional story about a billionaire paying his way into space began in 2018, when Yusaku Maezawa, the founder of ZOZO, Japan's largest online fashion retailer, announced his intentions to fly to the moon on Musk's Starship during a joint event with the SpaceX CEO.

Dubbed "dearMoon," the effort aims to send artists and other creatives into space to see what beautiful things might come from bringing people with uncommon perceptions to the final frontier. "What if Basquiat had gone to space?" asked Maezawa during the 2018 event, unprompted, which streamed on SpaceX's YouTube channel. "Once I got started, I couldn't stop thinking of who else [might have experienced something]" like cosmic transcendence.

The main requirement, besides winning another lottery, was for each applicant to pitch their plan to create something imaginative, inspired by their journey into lunar orbit and back again. "I thought about how this can contribute to the world and world peace. It's my lifelong dream," said Maezawa. "These masterpieces will continue to inspire all of us." Honestly, he might be right.

But this was a pivot from Maezawa's initial position about his forthcoming trip to the moon. At first, his campaign called for the selection of an "intimate space partner" to accompany him on his lunar trip. It was exactly what it sounded like, and it seemed to be going well for him. The website had amassed 27,722 applicants according to an Interesting Engineering report. Plans were made for the Japan-based streaming service AbemaTV to document the mission via a reality show, dubbed "Full Moon Lovers." This was going to happen, and it felt like something from the mind of James Franco.

Later, when the show was canceled, it turned out Maezawa had called off his romantic search, citing "personal reasons," according to a tweet from the aspirational astronaut. With nearly 30,000 applicants, there were probably a lot of disappointed space companions. But they were given the opportunity to apply again in March 2021, when the Japanese billionaire offered the public a chance to win one of eight seats on SpaceX's Starship; he'd purchased the tickets for a private mission to the moon in 2023, he explained in a subsequent tweet.

By July, Maezawa had selected twenty finalists for his art project in space. "Coming close to the end of the selection process for dearMoon," wrote the Japanese fashion tycoon in an Instagram post on July 15. He later shared a YouTube video that included clips of applicants expressing varying degrees of noble ambitions for what they would do while in transit to the moon and back. The contestants represented an array of creative professions and crafts, from dancers and DJs to painters and photographers. Even Olympic gold medalists were in contention for one of the golden tickets.

"I would consider this to be the most ambitious and probably one of the greatest artistic collaborations ever," said Boris Moshenkov, a Vancouver artist who was among the finalists, in an Instagram video, according to a report from the *Observer*. "That's what gives me goosebumps every time I think about the project." Another contestant named Tracy Fanara, a scientist at the National Oceanic and Atmospheric Administration (NOAA), found the project too exciting to take lying down.

"I did not sleep for like six weeks going through the process," said Fanara in a *DailyMail* report. "Let's just say, going through the process and getting to each step is just crazy. To think you might actually be a part of something so much bigger than yourself."

No matter how much excitement each contestant expressed, or how hard

There's no avoiding the resemblance of the New Shepard rocket to other, phallic objects.

BLUE ORIGIN
BLUE ORIGIN

they tried to manifest a special cosmic destiny, only eight would be selected to join the billionaire. Together, the nine civilian passengers would be joined by a handful of SpaceX employees, bringing the crew complement up to ten or twelve.

Of course, no one from the dearMoon project will set foot on the moon. At least, not this time: The trajectory is a three-day journey, followed by a high-speed swoop around the dark side in less than one day. The return to Earth will go down three days later. In form, the flight plan isn't so different from the first missions to the moon during the Apollo program—those few missions before Apollo 11 made its historic landing. Apollo 13 also made a similar flight, but hopefully, dearMoon's mission won't suffer any potentially life-threatening incidents.

↑ A lot of people wanted to go with Maezawa to space. Matezawa is, fittingly, very charismatic.

→ Once the dearMoon project gets off the ground, old gems like this Apollo image will get a major update.

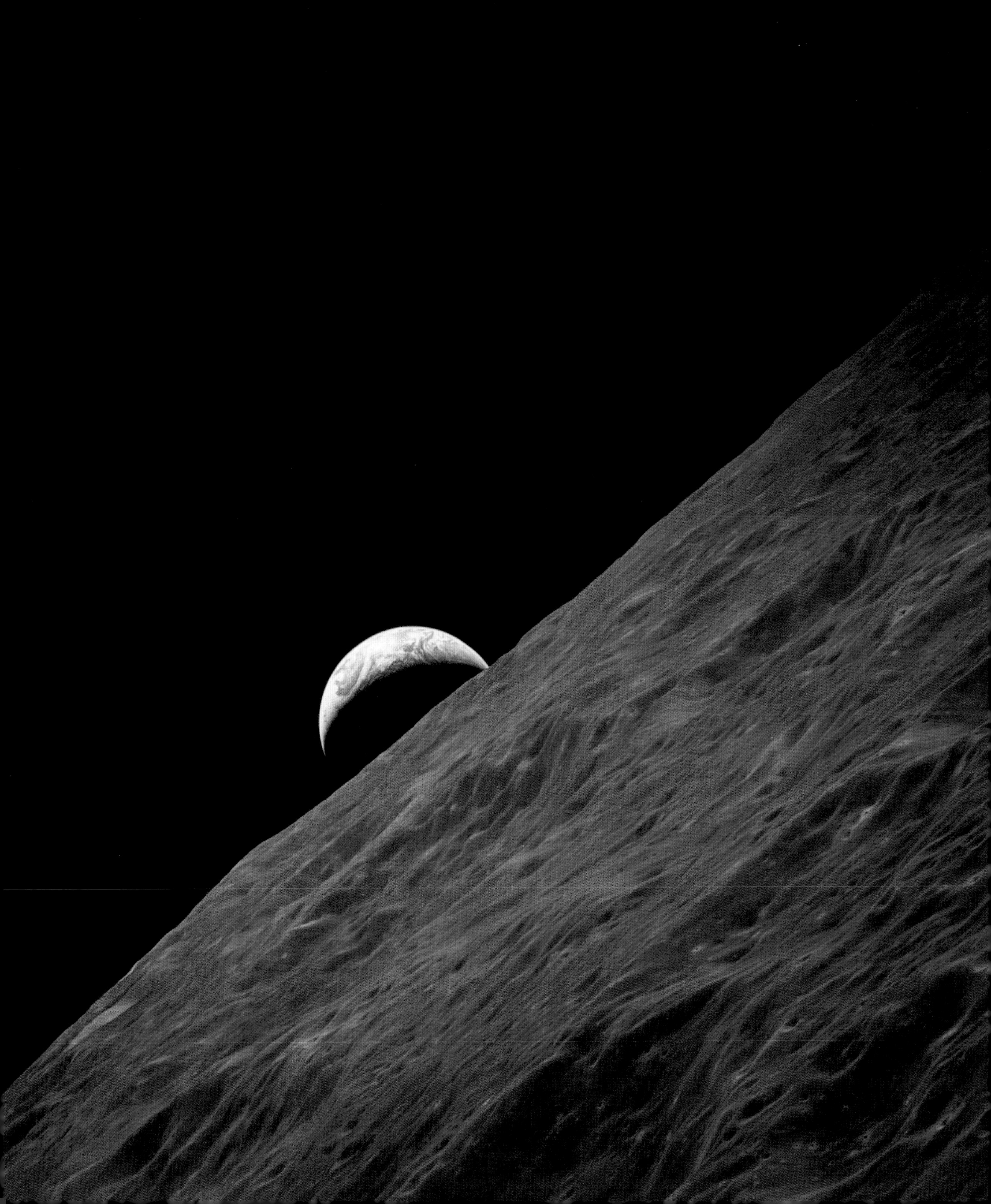

中国载人航天
中国航天

7

TROUBLE ON THE MOON AND MARS—AND EARTH

Until very recently, it seemed a foregone conclusion that the second Space Race would be a friendly, if at times rude, rivalry between major aerospace firms vying for contracts with NASA and other major space agencies. But in the last several years, a new power has rapidly accelerated its own growth into space exploration. That new power is the Middle Kingdom: China. And, it's committed to becoming a "great space power," said the nation's president, Xi Jinping, according to a report from the *New York Times*.

While China has nurtured an interest in developing a human space program since the 1950s and 1960s, reportedly even selecting nineteen astronauts for the Shuguang-2 program before it was canceled in 1972, the 1980s and 1990s saw a renewed commitment eventually get off the ground, when the Shenzhou program successfully launched a robotic mission on November 20, 1999. Then, came China's first crewed launch: the Shenzhou 5, on October 15, 2003, making it the third nation to put humans in space.

The launch of a Chinese Long March 5B rocket

These missions went forward under the purview of the China Manned Space Program (CMS), which continues to launch humans into space. After this success, China set its sights on matching the accomplishments of the former Soviet Union's Mir space station and the United States' Skylab (and now International Space Station) by launching its own space station: Tiangong-1.

China's first space station was launched on September 11, 2011, atop a Long March 2F rocket from the country's launch facilities in its northwest region. The orbital station's name translates to "Heavenly Palace 1," and it weighed 9.4 tons (8.5 metric tons). It was nearly 34 feet (10 m) long and 11 feet (3 m) wide and housed a single experiment module, along with a resource module to maintain propellant tanks and rocket engines.

Its initial orbit was 217 miles (349 km) high, slightly closer to Earth's surface than the ISS. China's first station was powered via two solar arrays, and it could support three astronauts at a time. One of its central roles in China's space program was to serve as a platform for space-docking procedures, a crucial capability for any space power interested in expanding its presence to enticing destinations beyond Earth orbit, like the moon and Mars.

China launched a second space station called Tiangong-2 on September 15, 2016, to take its technological prowess in low-Earth orbit even further. Shenzhou 11, a crewed docking mission, traveled to the new station in October. Later, in April 20, 2017, a new cargo ship dubbed the Tianzhou-1 was launched atop a Long March-7 Y2 carrier rocket. After two days of careful guidance, the Tianzhou-1 cargo vehicle successfully docked with the Tiangong-2 space lab. There were three docking iterations, the last of which was successfully completed in only six and a half hours, far more efficiently than an initial two-day iteration, according to a *GBTimes* report. This rapid jump in space maneuvering abilities would foreshadow how quickly China would become a major factor in the second Space Race.

But before progress could begin in earnest, China, too, suffered a major setback. On April 1, 2018, Tiangong-1 was ripped apart by the fiery blaze of reentry plasma as it streaked through the skies above the Pacific Ocean at roughly 8:16 p.m., according to the Joint Force Space Component Command (JFSCC) of U.S. Strategic Command. "The JFSCC used the Space Surveillance Network sensors and their orbital analysis system to confirm Tiangong-1's reentry," read the Air Force's statement confirming the hellish end of China's first space station. Tiangong-1's operational life was just two years,

← The interior of China's Tiangong-1

→ A space module docking with China's orbital station

→ Mission control at China's Jiuquan Satellite Launch Centre

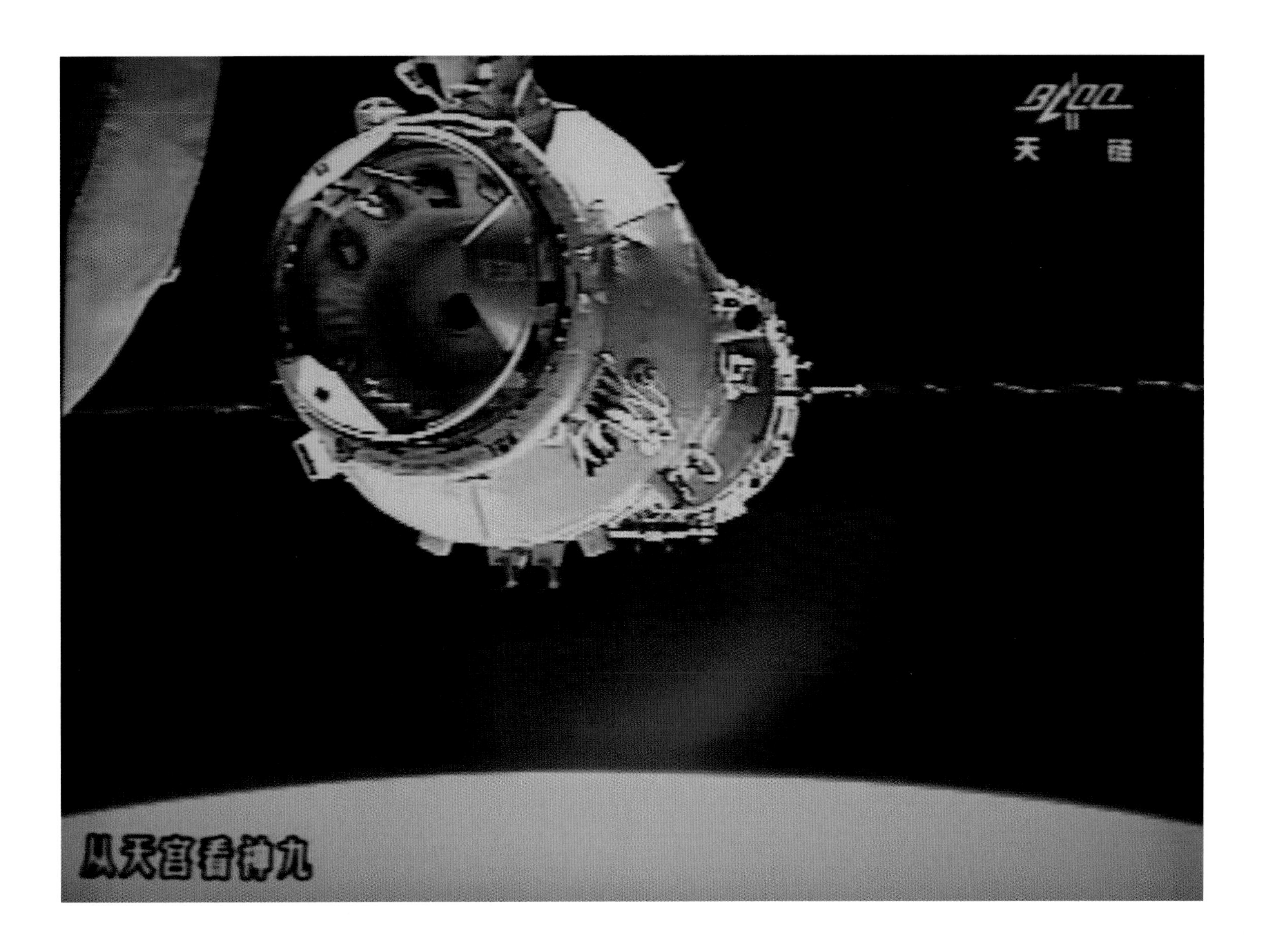

but its functional mission profile was basically completed by the time it reached its end.

But it wasn't all pulverized in the atmosphere; some fragments of the bus-sized Tiangong-1 probably survived reentry and plunged into the depths of the Pacific; hardly anyone was at risk, with a one-in-one-trillion chance of being smacked by debris from China's now-defunct Heavenly Palace, according to a *Space.com* report. But if anyone did find a chunk of the destroyed station, they might not be well: Experts have said that space junk could be contaminated with hydrazine, an extremely toxic rocket fuel.

CZ-3B

China's Long March 3B launching

A Reddit user in Guam reportedly found dangerous "honeycomb structures" in debris from China's rocket.

Speaking of toxic space junk, this may remain a major theme for China's space program, to a much greater extent than the debris that concerns residents near the launch sites of SpaceX or other U.S. aerospace firms. On November 29, 2019, China launched a Long March 3B carrier rocket into orbit, and it left a path of destruction in its wake. Lifting off from the Xichang Satellite Launch Centre at 7:55 p.m. EST, the mission successfully delivered its two payload satellites into orbit—the Beidou-3 M21 and M22—to an average altitude of roughly 13,545 miles (21,799 km). But not everything was up to snuff. Shortly after launch, one of the lower-stage rocket boosters dropped into someone's house without warning or even an official follow-up report, according to a tweet containing a video post from Weibo, the nation's proprietary social media alternative to Twitter. Yellow smoke could be seen pouring out of the ruined structure—smoke came from highly toxic hypergolic propellant.

It remains unclear if anyone was harmed from the impact, but the Chinese government had given notice to people living in the drop zones suggesting that they evacuate during the launch. Residents were also told to stay away from any wreckage that showed up in their neighborhoods, since harmful chemicals are used during launches. On the upside, China reportedly compensates residents whose property is damaged by raining rockets.

Early in April 2020, a rocket from China failed mid-launch procedures while lifting an Indonesian communications satellite into orbit, and a big chunk of it might have fallen in Guam, about 3,100 miles (4,989 km) from the launch site at Xichang Satellite Launch Centre, according to a Reddit user. The wreckage, which was almost ten feet (3 m) wide, may have been carcinogenic. Bystanders in Guam reportedly saw the fiery debris streak across the moonlit sky. Guam's Offices of Homeland Security and Civil Defense made an official statement that connected the debris to China's downed rocket, adding that there was "no direct threat" to the Pacific Islands, according to a *Spaceflight Now* report.

Initially, suspicions surrounding the wreckage hinted that it might be part of China's rocket from between the oxygen and hydrogen tanks. But another comment on Reddit from an aerospace engineer claimed that "honeycomb structures are very common even in parts where you would not expect it . . . Honeycomb is kind of the ductape [*sic*] of aerospace construction." The wreckage, it turned out, could have been from the first or second stage

↑ This plasma-like substance shows up when any large-enough body reenters Earth's atmosphere, and we can see it from the surface.

→ China's "taikonauts" now give aspirational talks with students on the ground.

of the rocket and possibly "incredibly carcinogenic," said another Reddit post.

"If the honeycomb structure is carbon fib[er], inhaling those tiny fib[ers] can't be good for you," added the Reddit post. This was worrying not because each successive rocket fall endangered another local community with potentially carcinogenic chemicals, but because it didn't look like China took the incidents seriously—like it doesn't matter what damage their space program caused, so long as the space missions continued.

On May 11, 2020, one of the largest chunks of space debris ever recorded made an uncontrolled reentry into Earth's atmosphere, splashing down into the Atlantic Ocean near the northwest African coast. Also part of a Long March 5B rocket, the debris was from a mission that successfully launched a payload into orbit from Wenchang Space Launch Center in South Hainan province, China. But after its successful launch, several tense hours ensued.

The chunk of rocket wreckage weighed nearly 20 tons (18 metric tons), according to a *CNN* report, making it of significant concern to any vulnerable structures or people beneath. "At 17.8 tonnes, it is the most massive object to make an uncontrolled reentry since the 39-tonne Salyut-7 in 1991, unless you count OV-102 Columbia in 2003," tweeted astronomer Jonathan McDowell of the Harvard-Smithsonian Center for Astrophysics.

The U.S. Air Force's 18th Space Control Squadron later confirmed the incident, tweeting, "#18SPCS has confirmed the reentry of the CZ-5B R/B (#45601, 2020-027C) at 8:33 PDT on 11 May, over the Atlantic Ocean. The

#CZ5B launched China's test crew capsule on 5 May 2020."

When it comes to rocket debris, the bigger it is, the more likely it is to survive reentry and smash into Earth, McDowell told *CNN*: "Once they reach the lower atmosphere, they are traveling relatively slowly, so worst case is they could take out a house."

It's especially difficult to predict where space debris will land if it's plunging into the atmosphere in an uncontrolled descent, which also limits what we can know about potential damage on the planet below. "For future reference note that DoD's announcement [of the falling rocket debris'] reentry time came about 1.5 hours after the event—by historical standards pretty fast," added McDowell in a tweet. "In the absence of eyewitness reports, the info always comes with a delay which is why I kept saying 'if the thing is still up,'" McDowell explained of the inherent unpredictability of rapid descents into Earth's atmosphere.

For the May 2020 incident, SpaceTrack, which monitors objects in orbit, could merely predict potential crash sites of the debris in Australia, the United States, and Africa. That's a pretty wide window. "The problem is that it is traveling very fast horizontally through the atmosphere and it's hard to predict when it will finally come down," McDowell said in another Twitter post. "The Air Force's final prediction was plus or minus half an hour, during which time it went [three-quarters] of the way around the world.

"It's pretty hard to do any better," McDowell added.

China's Space Program Hits Its Stride

No humans were harmed by the May rocket debris from China's launch, but this crisis would not be the last. A year later, on May 8, 2021, astronomers with the Virtual Telescope Project captured an image of China's Long March 5B rocket tumbling into the atmosphere in what was the largest-yet uncontrolled reentry ever recorded.

The telescope's robotic unit, called "Elena," monitored the rocket as it arced across the night sky at roughly 0.3 degrees per second. "At the imaging time, the rocket stage was at about 700 km [435 miles] from our telescope, while the sun was just a few degrees below the horizon, so the sky was incredibly bright[.] These conditions made the imaging quite extreme, but our robotic

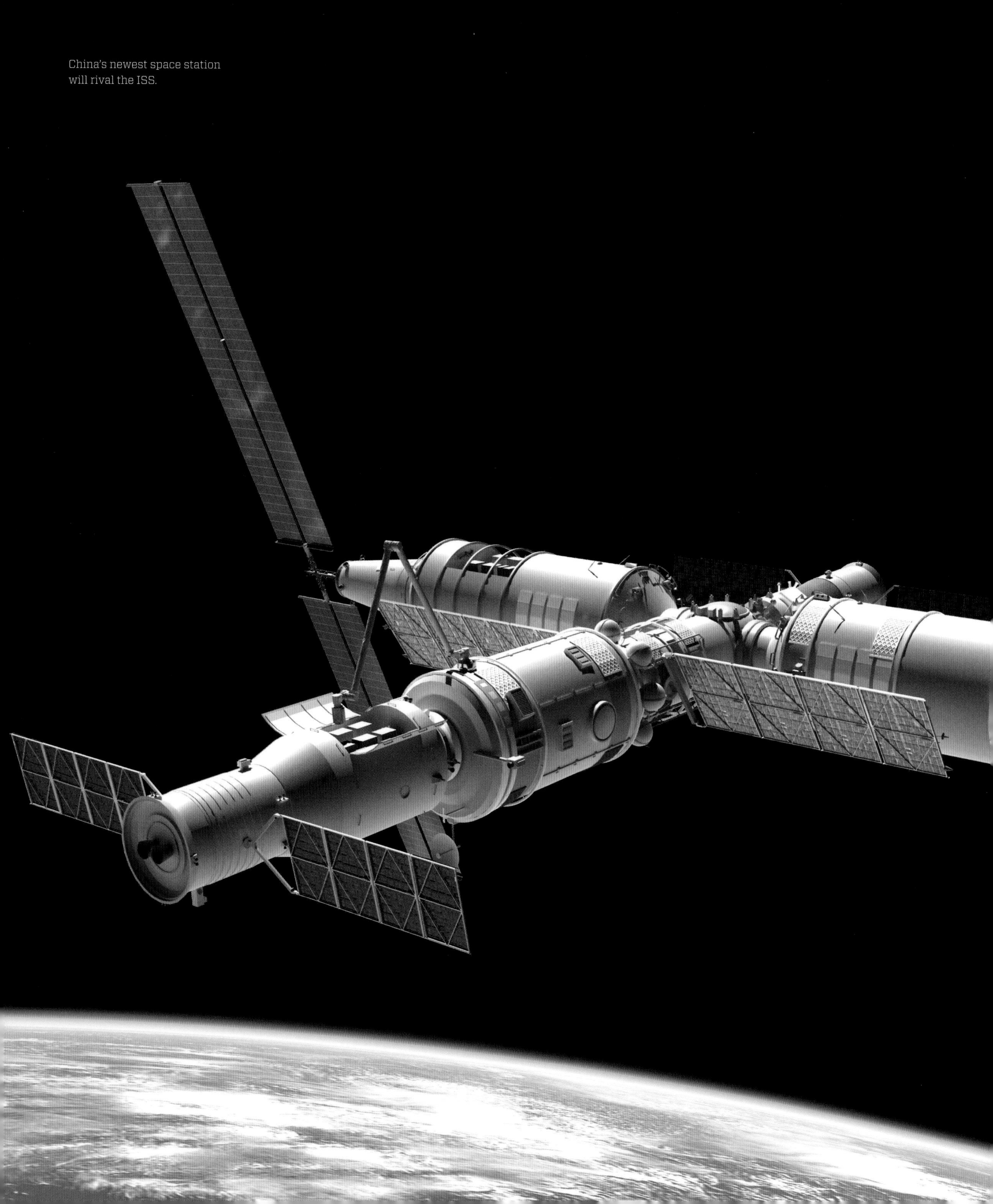

China's newest space station
will rival the ISS.

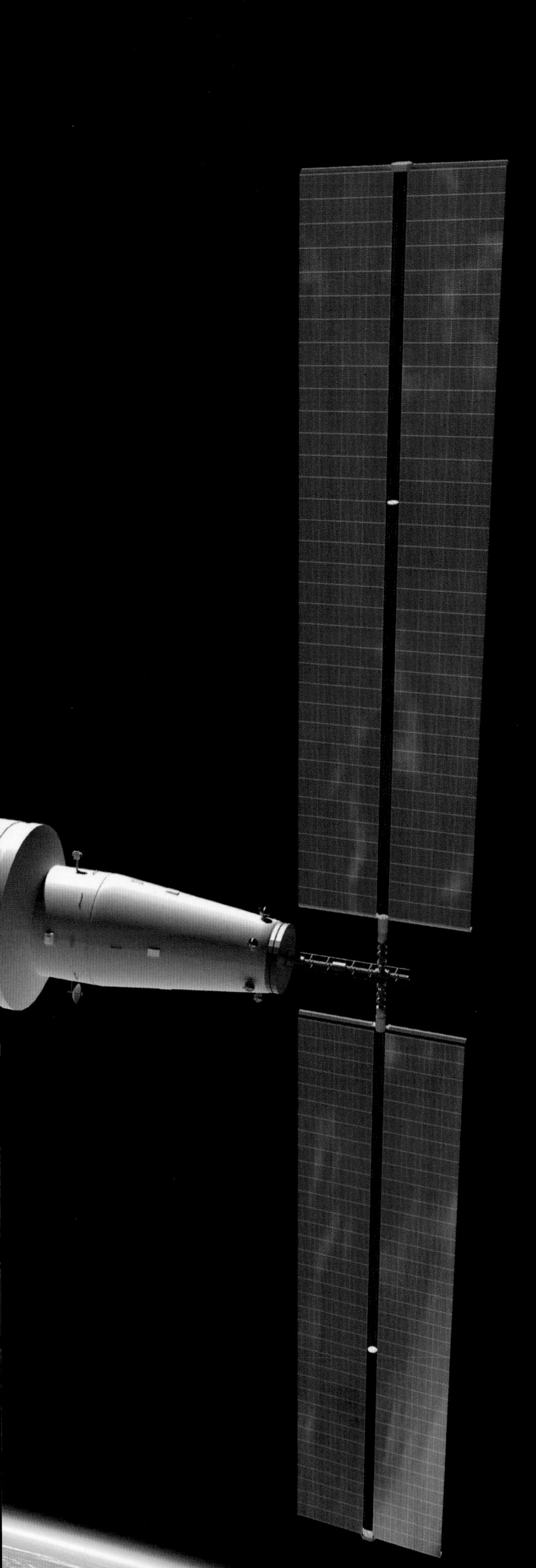

telescope succeeded in capturing this huge debris," said Gianluca Masi, one of the astronomers attached to the Virtual Telescope Project, according to a Space.com report.

While the tumbling rocket looked like a tiny dot in the image, China's Long March 5B was ninety-eight feet (30 m) tall—no small speck, by any estimation. And this speck of light, portentous as it was, also represented a new beginning for China: the tumbling rocket had just delivered the first of eleven parts for the nation's new space station into orbit. Called the CSS (China Space Station), the completed station was slated to become fully operational sometime in 2022. But this May 2021 launch had lifted only the station's living quarters, called the Tianhe, into orbit.

This orbital insertion was successful, but, critically, the Long March 5B rocket that put it here quickly began one of the most concerning uncontrolled reentries ever witnessed. This looked like a signal to the world from China that its space ambitions were more important than not only the livelihood of other nations' space ambitions, but also the lives of their constituent populations. U.S. White House press secretary Jen Psaki declared a need for "responsible space behaviors" at a press briefing, while McDowell, the Harvard astronomer, described China's space logistics as "negligent," a loose process that "makes the Chinese rocket designers look lazy."

The U.S. space programs, by contrast, haven't allowed objects weighing more than ten tons (9.1 metric tons) to undergo uncontrolled reentry since 1990. From a certain point of view, we could argue that the blunders of China's budding space program mirror those of the United States and its allies during their salad days in space. But whether via shaky beginnings or callous ambition, China's space station brings it a step closer to rivaling Western space efforts.

Specifically, China's new space station will rival the International Space Station, and subsequent launches with additional components and modules have gone without incidents like uncontrolled reentries. On October 15, 2021, China launched the Shenzhou 13 via a Long March 2F, lifting three Chinese nationals—Zhai Zhigang, Wang Yaping, and Ye Guangfu—to the core module of the recently deployed Tiangong space station. This was the sixth of eleven missions to install the new station.

As of early 2022, the last successful launches had happened on December 29, 2021, when China closed out a record-breaking year of launches with two orbital launches lifting off from the Xichang and Jiuquan spaceports hours apart. The former was a Long March 3B rocket that launched from complex 2 at the Xichang spaceport at about 11:43 a.m. Stowed aboard was a new communications test satellite (TJSW-9), which was placed in geosynchronous orbit. The China Academy of Space Technology (CAST) manufactured the TJSW-9, but the lack of public records suggests that it could be a military satellite.

Hours before TJSW-9, the Tianhui-4 mapping satellite launched atop a Long March 2D at 6:13 a.m., achieving orbit. By the time news of both launches reached the United States, space tracking systems had already cataloged the two new orbital objects.

China launching a space satellite from its Xichang spaceport

Z-3B

中国航天

← A launch from China's Jiuquan spaceport

→ China's first female taikonaut (astronaut), Liu Yang

This brought an incredibly busy year of launches for China to a close, leaping high above its previous record of thirty-nine in 2018 and 2020. December's final launches were China's fifty-fourth and fifty-fifth, but perhaps more significant were the country's claims of a hypersonic vehicle test, initially reported by the *Financial Times* in October 2021.

Space War Tactics

Sometime in summer 2021, China successfully tested a hypersonic missile that "went around the world," according to the second-highest U.S. general, reported *CNN*. "They launched a long-range missile," said General John Hyten, an outgoing vice chairman of the Joint Chiefs of Staff. "It went around the world, dropped off a hypersonic glide vehicle that glided all the way back to China, [and then] impacted a target in China."

Much like the first Space Race, this event revealed that the international dynamic of Space Race 2.0 would likewise serve as an extended domain for mounting tensions between rival nations. Hypersonic missiles are a strategic first-strike weapon that override traditional fail-safes against conventional ballistic missiles. In the old days of the Cold War between the Soviet Union and the United States, active war (and the implication of nuclear war) was kept at bay in part because any nuclear missile would follow a high arc through the upper atmosphere, giving defensive radar enough warning before detonation to allow the defending nation to launch its own nuclear arsenal.

With hypersonic missiles, that warning may never come. Instead of achieving supersonic speeds by rocketing to the edge of Earth's atmosphere and back down, hypersonic missiles

↑ Hypersonic weapons tests have increased tensions for the USA and its Asia-based allies.

↓ A Russian ballistic missile exiting its silo

rapidly accelerate to several times the speed of sound and reach their targets in a very short time, and—crucially—without showing up on radar in time to mount a defense, let alone a response. Disturbingly, hypersonic missiles can carry nuclear weapons. As of early 2022, it looked like China was the only nation with fully functional hypersonic missiles. And we know very little about how much progress the country has made since the summer of 2021.

This isn't to say there aren't ways of defending against a hypersonic missile attack, but it requires state-of-the-art real-time tracking technology, synchronizing on-site radar with satellite-based tracking systems, conjoined in parallel to provide next-gen point-defense or intercepting weapons with split-second decision-making capabilities. When seconds or less make the difference between averting an attack or taking critical damage, every link in the defense network is absolutely crucial. This means that an aggressor with hypersonic missiles could effectively "blind" an enemy's defense infrastructure by preemptively knocking out military satellites.

That's not a new idea. It was initially popularized during the Reagan administration under the "Star Wars" program of the 1980s. And this idea of turning low-Earth orbit into another battlefield is alive and well: On November 16, 2021, Russia executed an anti-satellite weapons test, obliterating one of its own orbital satellites and generating more than 1,500 chunks of

It would be tragic if the space between the Earth and the moon became a battlefield.

supersonic space debris that placed the International Space Station—along with the lives of seven astronauts—in danger.

When it happened, four U.S. astronauts, two Russian cosmonauts, and one German astronaut were aboard the ISS. Upon learning of the imminent danger, they rushed to take shelter in docked capsules to reduce the risk to their lives should one of the thousands of space debris chunks penetrate the hull of the fragile station. The space junk created from Russia's destroyed satellite were massive enough to show up on orbital radar and continued to pose a periodic danger to the ISS for days. "It was dangerous. It was reckless," said the U.S. State Department's Ned Price in an official press briefing. "It was irresponsible."

And, it was actually worse than this: Beyond the initial 1,500 detectable fragments were countless more too tiny to monitor. Even these miniscule bits carried sufficient momentum to damage the ISS and other orbiting satellites. "We are going to continue to make [it] very clear that we won't tolerate this kind of activity," Price added in the AP News report.

One of the NASA astronauts aboard the ISS during the ordeal, Mark Vande Hei, told AP it was "a crazy but well-coordinated day" before falling asleep. "It was certainly a great way to bond as a crew, starting off with our very first workday in space." Meanwhile, the U.S. Space Command monitored the tactical nightmare from the surface. "Russia conducted an anti-satellite missile test," tweeted U.S. Space Command. "Russia continues to weaponize space. [We stand] ready to protect/defend US/allied interests from aggression in the space domain."

In response to a surge of international concern, Russia's space agency, Roscosmos, officially confirmed the incident. "The Space Station crew is routinely performing operations according to the flight program," the agency tweeted, according to an Interesting Engineering report. "The orbit of the object, which forced the crew today to move into spacecraft according to standard procedures, has moved away from the ISS orbit. The station is in the green zone." This was only half-true, since the nature of orbital bodies is that they return to roughly the same relative location with every revolution, in an event called a "transit."

"A few minutes away from the next debris field transit for ISS," tweeted McDowell, the renowned astronomer from Harvard's Center for Astrophysics. Needless to say, he was not pleased

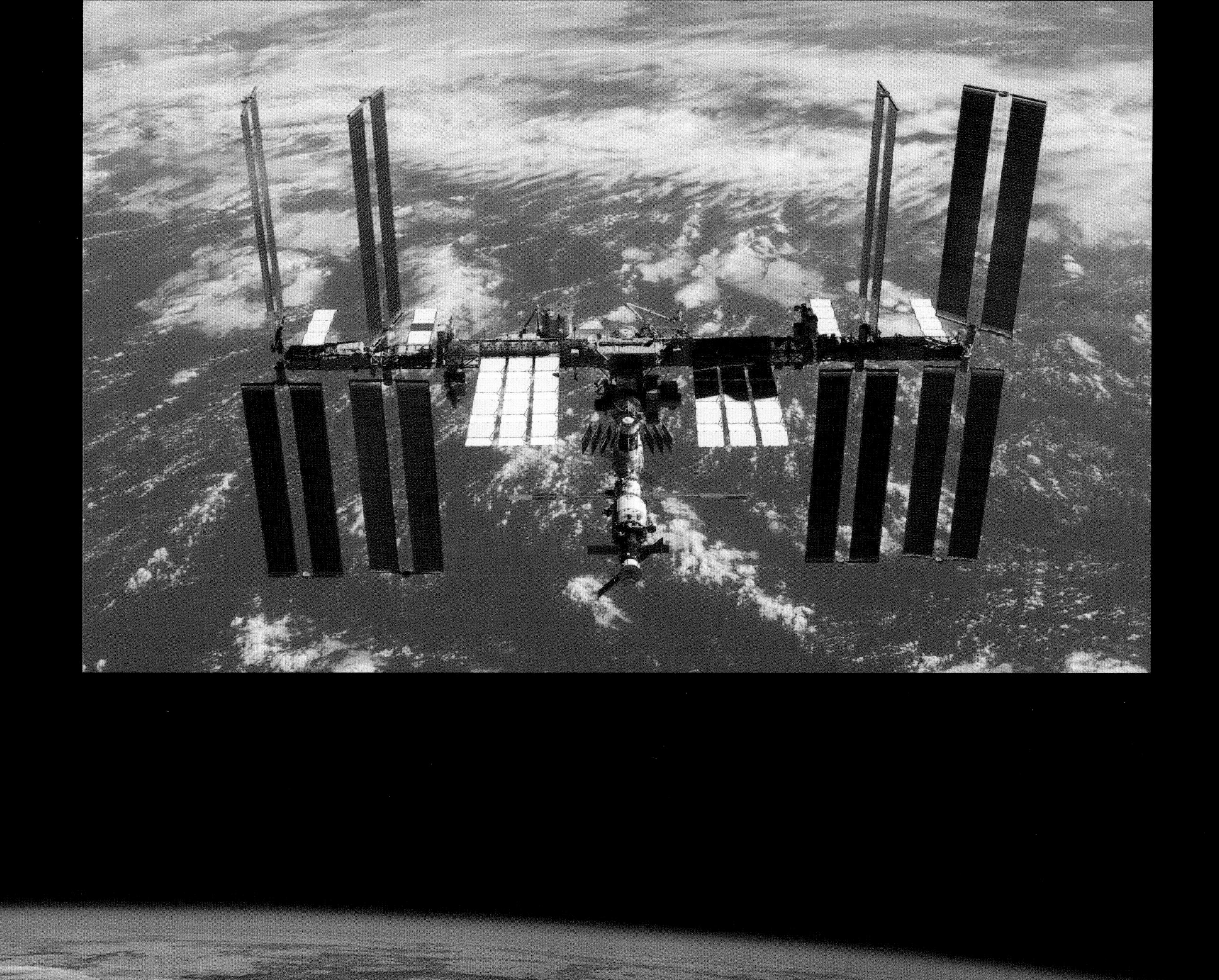

← The ISS still needs continual work to keep it functioning in orbit, independent of the second Space Race.

→ Astronaut Kayla Barron smiles while peering out from a window inside the ISS's cupola. Eventually, the ISS will be put into a deorbit burn and die in a great ball of fire.

with the prospect of anti-satellite weapons tests in low-Earth orbit, where human space operations are at their most precarious. "I condemned the 2007 Chinese test, the 2008 US test, the 2019 Indian test, and I equally condemn this one," added McDowell in a follow-up tweet. "Debris-generating antisatellite [*sic*] tests are a bad idea and should never be carried out."

Sadly, these words of warning will likely go unheeded by both Russia and China. Neither government subscribes to the same level of relative openness and "freedom of the press" regarding their space activities as the United States or Western Europe. But there's a flip side: Among other things, this means that even when China or Russia praise their missions or technology as 100 percent successful, we should take it with a grain of salt—they have every incentive to appear more powerful and advanced than they actually are.

A New Challenger on the Moon

It's safe to say that China's increasing presence in space since the Tiangong-1 space station has reached full swing. While there's nothing wrong with international competition expressed through a Space Race (it happened before with the now-defunct Soviet Union), concerns have been repeatedly raised about the consistent danger China's launches pose to Earth-based communities and environments. A main rocket booster of the Long March 5B fell into an "uncontrolled reentry," smashing into the Indian Ocean in May 2021, nearly impacting the Maldives.

By June, the first Tiangong crew had begun living and working in orbit after several more modules were delivered, in addition to supplies. But this wasn't the first Chinese space station: Two short-term prototypes have gone up and down, but once Tiangong is complete, it could last more than a decade. That's

↓ A Mars probe is launched on a Long March-5 rocket from the Wenchang Space Launch Site. We should expect China to increase its annual launches substantially, but it's unlikely the nation will keep pace with the rapidly multiplying number of launches operated by public-private aerospace partnerships in the United States.

→ China will continue to send taikonauts to its space station, much like the USA and other nations have to the ISS.

significant because it could outlast the ISS, which NASA has said might be retired and put through a controlled reentry demolition in the next decade, although more recent plans hint that the U.S. agency wants to make it last longer.

And as of January 2022, China planned on more than forty launches during the year. China's forthcoming launches will loft two Shenzhou crewed missions, in addition to two Tianzhou cargo spacecraft launches and the installation of the station's next two modules, according to an AP News report.

This is significant not only because China is beginning to keep pace with NASA, but also because it has a good chance of surpassing NASA (not counting the combined missions of public-private partnerships). The ISS, which was jointly developed by the United States and several European nations (including Russia), is approaching the end of its life, which was initially slated for 2024. If it's fully decommissioned, that would make China the only major power with a permanent human presence in low-Earth orbit. And soon, its presence could extend even farther. In 2019, China became the first nation to land a probe on the moon's far side, which forever faces away from Earth. It was the country's second moon landing, after its first, in 2013.

And China's latest rover continues to operate, despite being far beyond its initial three-month life span. On September 29, 2021, it had been operational for 1,000 days, having moved more than half of one mile (839 m) from its landing point in the Von Kármán Crater, near the lunar south pole, according to a statement from the project directors. In December 2020, China had sent a different spacecraft to the moon to scoop nearly four pounds (1.8 kg) of rocks and soil close to a volcanic region known as Mons Rümker and then returned the sample to Earth.

↑ We might soon live in a world where only China carries out low-Earth orbit space walks. But even if there's a gap for the USA and its private aerospace partners, it won't last long.

This was the first lunar sample returned since the now-defunct Soviet Union's Luna 24 mission in 1976. China's latest moon probe was dubbed Chang'e, a moon goddess from Chinese mythology. Three additional moon probes are slated to reach the moon by 2027, including more rovers, a flying probe, and even one that might attempt 3D printing in space, according to a report from the *New York Times*.

And these highly ambitious lunar missions are intended to lay a new interplanetary foundation for future moon bases to be visited by taikonauts (Chinese astronauts), sometime in the 2030s. And, there's more. In spring 2021, China signed an agreement with Russia to design and construct an International Lunar Research Station (ILRS) orbiting the moon, according to statements made by both countries' space agencies. This breakthrough came after months of talks between the two nations.

Russia had considered joining NASA's Gateway program, which would give it one voice among many in a colossal coalition with private aerospace firms like SpaceX to build the next generation of deep-space exploration. But Russia declined the opportunity, opting instead to build "a complex of experimental research facilities created on the surface and/or in orbit of the moon," according to a Roscosmos statement reported in *The Verge*. The two-party lunar project, when completed, could provide a platform for a broad variety of research

↑ China plans to build a moon base one day, likely in collaboration with Russia's Roscosmos.

← A model of an anti-satellite weapon is displayed at a Republic Day parade in New Delhi. Anti-satellite weapons tests put everyone in jeopardy and risk undoing decades of work in space.

← China had already put two landers on the moon by 2022, one of which, the Chang'e, was the first ever placed on the moon's dark side.

↓ China even landed its first Martian rover in 2021, the second nation to do so (at least, successfully).

experiments "with the possibility of long-term unmanned operation with the prospect of a human presence on the moon."

Chief of Russia's Roscosmos, Dmitry Rogozin, and head of the China National Space Administration (CNSA), Zhang Kejian, jointly signed the new lunar station agreement digitally. While from these appearances it seems like China and Russia are actively seeding a renewed Space Race, the agreement was a necessity for China's space ambitions. This is because, in 2011, Congress passed the Wolf Amendment, which prohibits NASA from working with China. But for Russia, it represents the latest step away from collaborative projects with the United States going all the way back to 1963, when President John F. Kennedy proposed that the United States partner with the Soviet Union on a joint lunar mission.

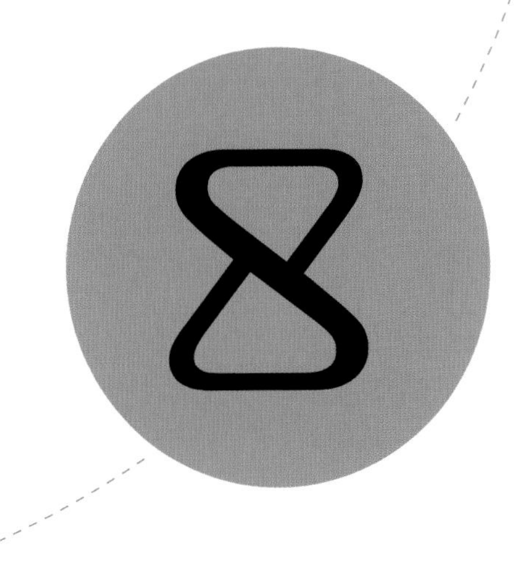

THE FUTURE: CONFLICTING REALITIES

That the very idea of a joint mission to the moon between Russia and the United States, let alone China and the United States, feels like a bizarre paradigm is the consequence of decades of complicated geopolitics and a move in the United States away from partnerships with new space powers. The necessity for international partnerships was more than goodwill or a drive to peace. Surprisingly, it's linked inextricably to the history of colonialism. Head of Blue Origin Jeff Bezos has not been unambiguous about his intentions to colonize the entire solar system, justifying this expansion with promises of a new level of prosperity fueled by boundless resources, like precious metals in asteroids, lying in wait.

Thanks to Bezos, Musk, and Branson, we may see signs of human life on the moon before the 21st century comes to a close. But the human cost could be high.

Luckily, there are no known intelligent life-forms besides humans in this solar system, which means that, as Bezos and other space barons expand their industrial empires into the solar system, no living beings will be subjugated. But to extract the resources from asteroids, the moon, and Mars, the space barons will need human labor. And, if allegations against worker relations at

Many claim conditions in Amazon facilities are grueling. One shudders to imagine how difficult work environments would translate to space, on the moon or Mars, where no one is coming to help you.

Bezos's and Musk's companies are true, then work in space could be a grueling and nightmarish enterprise.

In September 2021, twenty-one present and former Blue Origin employees published an open letter wherein they allege the firm's work environment is abusive, prejudiced, and "toxic." They went on to argue that resistance was truly futile, as safety concerns were passed over, causing some to experience suicidal ideations. "One directive held out SpaceX as a model, in that 'burnout was part of their labor strategy,'" read the open letter. "Requests by managers and employees for additional engineers, staff, or spending were frequently denied, despite the fact that Blue Origin has one of the largest single sources of private funding on Earth."

The open letter also claimed that workers were consistently told to "be careful with Jeff's money," and were discouraged from asking for additional funding or resources, and to "be grateful." Musk's company, SpaceX, was sometimes called "SlaveX" by employees, according to a job board review of SpaceX titled "Called slavex for a reason." While the pay is great, and overtime was offered, it seems management refused transfers, and sometimes employees were told without warning that they needed to work six-day weeks for the next two months. "If you're all about money[,] great," read the review. "[B]ut how are you supposed to enjoy it if you're working your life away."

So much for preserving "the light of consciousness." Even if there is only a small chance of workers finding better, more ethical employment elsewhere, on Earth they can at least step outside and breathe free air. On a lunar- or Mars-based colony, there's no guaranteed ride home if the gig doesn't work out. Musk has said he wants "direct democracy" on Mars, but how can the people rule when the very technological basis for their daily lives is owned and operated by a private company? "Vote against the boss? No rations for you. Labor unrest? Try striking without oxygen," read a 2022 essay in *The Baffler,* by Corey Pein, on the dangers of privatized space colonies.

On January 21, 2022, creator of a software program called The Guide that tracks asteroids, comets, and generic near-Earth objects announced that one of SpaceX's Falcon 9 rockets had entered a terminal trajectory with the moon—and it was slated for impact on the lunar surface on March 4. No one would die from "this moonfall," and no science missions or lunar probes were on the line, but the tracking data predicted "certain impact," said Bill Gray.

However, on February 12, Gray published a correction to his initial prediction: Upon completing additional research, he realized the object on course to punch the moon was really debris from China's Chang-e 5-T1, the nation's predecessor to its highly successful Chang-e 5 mission, which returned a sample of moon rocks to our planet in 2020. A group of students from the University of Arizona confirmed Gray's updated conclusion, using a telescope mounted atop the college's Kuiper Space Sciences facility.

Notably, the students made their observations before Gray corrected himself—on January 21 and February 7. "They estimate that it will hit somewhere in or near the Hertzsprung crater on the moon's far side," said the university in a blog post. Whether from SpaceX or China, the end result would be the same: No one from Earth could see the moon get slammed by detritus, which consisted of a moon-bound rocket stage. This is because, moving at 1.6 miles per second (2.6 km/s), "the bulk of the moon is in the way, and even if it were on the near side, the impact occurs a couple of days after New Moon," Gray wrote in his original post. In other words, China's space junk would kick a darker region of the moon, left in the shadows by the lunar phases.

While some considered this imminent impact "no big deal," it sets a new precedent as the first substantial mass of space junk left on the moon (not counting the landing stages of the Apollo missions), which raises the question: How much space junk is too much? It's doubtful that a massive landfill will form on the moon, especially once settlements expand and astronauts recycle expired materials to save the rocket fuel it takes to send new hardware and metal from Earth. But as the second Space Race picks up steam in the coming decades, will the space barons or China mind terribly if large masses of trashed hardware are strewn across alien worlds?

We really need only look at Bezos's own words to grasp the basic utility space has for space barons. "When you look at the planet, there are no borders," he said in a July 2021 interview with *NBC News*. "It's one planet, and we share it and it's fragile. We live on this beautiful planet." Bezos had just become the second billionaire to reach space, and wearing his new cowboy hat, the ex-CEO of Amazon claimed he experienced what's called the "overview effect" during his flight, which is where astronauts feel a sense of interconnectedness with the vital smallness and fragility of Earth, which feels inherently borderless and unified.

According to Bezos, this experience inspired him to say that we should finally stop polluting Earth. Instead, he argued, we should move human industry off-world and pollute space. "You can't imagine how thin the atmosphere is when you see it from space," added Bezos in the *NBC News* report. "We live in it and it looks so big. It feels like this atmosphere is huge and we can disregard it and treat it poorly." The correct way to proceed might feel like a total reassessment of what it means to be human, whether the profit motive is really enlightened enough to make us an interplanetary species, and whether the quality of human life on Earth should come before the reach of a space baron's empire.

Alas, this was not Bezos's interpretation. "We need to take all heavy industry, all polluting industry and move it into space," he said in the report. It's great that Bezos seems to grasp, at least on a basic level, that Earth can't support the infinite growth of global industries that continue to alter the atmosphere, at least, not without endangering most forms of life, including humans. But moving industry into space involves more than "manpower" and energy-generating machines. Like Musk and space tourism firms have suggested, to expand human presence in space, we need to focus on the machines that build the machines—in other words: assembly lines.

But assembly lines are finely tuned machines specifically tailored to work with humans under full gravity, which means we're going to have to rethink on a fundamental level how we build things for a zero- or low-gravity environment. Then, we run into a different issue: Where will we collect the raw materials? There's no shortage of precious metals in asteroids, on the moon, and on Mars, but to put them to use for interplanetary industries, the space barons will have to implement a mining operation that amounts to an unprecedented and gigantic experiment of unspeakable proportions.

The technology to make it happen isn't quite here yet, but before we can start deploying the next Industrial Revolution beyond Earth, there's something else we're forgetting: Who has the right to claim territory in space, on the moon, or Mars? As of 2022, resource extraction beyond Earth's atmosphere is probably illegal. The Outer Space Treaty, which was presented to the UN Assembly in January 1967, unequivocally states that our entire solar system, and everyone within it, is the "Common Heritage of Mankind."

"We need to take all heavy industry, all polluting industry and move it into space"

—Jeff Bezos

This means the entire human race owns the solar system, like publicly owned land, and no one—neither person nor company—can stake an exclusive claim on cosmic bodies within it. This would mean that Jeff Bezos, Elon Musk, China, or anyone else would need to request permission from the entire world before installing extractive industries beyond the reaches of our atmosphere.

But this humanistic view of exploring higher realities in the universe for the sake of exploration doesn't vibe with entrepreneurs, who are, strictly speaking, interested in the bottom line. For the space barons, all extraterrestrial objects are potential financial resources and become the property of whoever can extract their raw material. This means if Bezos's company can build a vast mining operation to strip-mine the moon or Mars for profit, it will. And there are no signs that Musk wouldn't either.

The future isn't written in stone, but as things stand, the colonization of our solar system by the space barons might come to resemble previous colonial ventures on Earth, like the Dutch East India Company, which were licensed by colonial nations to extract and then sell any resources they found on the opposite side of the world, without the consent of the indigenous peoples. Perhaps anticipating this eventuality, the portent of a new era of space imperialism drove some former colonial states, including Pakistan and the Philippines, to build a treaty that more explicitly states how the solar system belongs to everyone, not just those who can afford to mine it.

Called the Moon Treaty, it states that space resources may only be mined with the approval of the entire world and ought to be shared equitably with the people of Earth. But this treaty has holes since none of the space-faring powers like the United States and Russia signed it when it was proposed in 1979. According to *Space.com*, space barons like Bezos even lobby against it. And, he's not doing it for the optics: In 2020, President Donald Trump signed an executive order that condemns the "Common Heritage of Mankind" principle of the Moon Treaty.

One of the most persuasive arguments for supporting the private colonization of planetary bodies and asteroids is the idea that taxed profits from mining raw minerals in space would substantially enrich the countries that license this mining or at least the nations where firms like Blue Origin are headquartered (in this case, the United States). But Bezos famously avoids paying taxes in the United States and worldwide.

Musk is hardly different, having only started paying full taxes on his holdings in Tesla in December 2021, worth more than $11 billion. Shortly after being declared *Time* magazine's "Person of the Year," Senator Elizabeth Warren stated the SpaceX and Tesla CEO was "freeloading off everyone else." Musk retorted that he'd be the largest taxpayer in U.S. history, according to a *Reuters* report. Musk doesn't collect a salary at Tesla but holds 22.9 million stock options slated for expiration in August 2022. In 2021, Musk could buy Tesla stock at $6.24 each, despite the market price fetching a much higher price of $900.

Additionally, the unconscionably wealthy can reduce tax burdens by making donations with unrealized gains (i.e., stocks) to nonprofits and charities. But there are ways for the space barons to do philanthropy without meaningfully supporting a humanitarian cause, effectively giving without giving. For example, Bezos gave up a free seat on his NS-19 spaceflight to Michael Strahan, a TV personality. That seat was worth $55 million, and it helped two people: Strahan and Bezos.

To add injury to insult, while the NS-19 space mission was underway, several dozen Amazon employees were trapped beneath a collapsed Amazon factory.

One shudders to imagine what horrors await future employees in the abyssal depths of outer space, where help, if it's coming, could be days, months, or even years away.

One chief feature of control is the ability to ignore the well-being of others. It's tempting to speculate (hyperbolically, in my opinion) that Musk wants to achieve fantastical interplanetary powers.

After all, he already declared himself "Technoking" of Tesla in a March 2021 filing with the Securities and Exchange Commission. Despite his uncommonly optimistic timelines for landing humans on Mars within the decade, it's very likely that neither he nor Bezos, nor any other present-day space baron, will live to see humans thrive on Mars.

Barring World War III or some other apocalypse, the very idea of landing humans on Mars has ceased to be a sci-fi indulgence and is transforming into a question of technology. While it's not a new question, we can at the very least ask whether we want to colonize the solar system so that the wealthiest few on Earth can enjoy the riches of the first interplanetary corporation, or find a way to reinvent society in space to forge a peaceful, egalitarian, and shared democratic legacy that will improve the human condition not only for those few who can afford it but for everyone who will ever live.

The technology to industrialize space isn't here yet, but there's something else we're forgetting: Who has the right to claim territory in space, on the moon, or on Mars?

AUTHOR BIO

Brad Bergan is a senior editor and cultural essayist working in online media. Previously, he was a contributing editor at *Futurism*, and his words have appeared in or on *VICE*, *The World Economic Forum*, the *National Book Critics Circle*, *3:AM Magazine*, and elsewhere. With investigative journalism cited in *Bloomberg*, *Discover*, and *NBC News*, he holds degrees in philosophy and English from the University of Iowa, and studied graduate-level creative writing at The New School. He lives in New York.

IMAGE CREDITS

B = bottom, **L** = left, **M** = middle, **R** = right, **T** = top

Alamy Stock Photos: 4, Geopix; 8, dpa picture alliance; 12B, Geisler Fotopress; 16, ZUMA Press; 18, UPI; 19, Jeff Gilbert; 20, Lucy Nicholson; 21T, Gene Blevins; 21B, ZUMA Press; 22, NTSB; 24, ZUMA Press; 26, Mariana Ianovska; 27L, Pictorial Press; 27R, stock imagery; 28–29, ZUMA Press; 34, McClatchy-Tribune; 36–37, NASA; 38T, dpa picture alliance; 38B, SpaceX; 39, SpaceX; 40T, Pierre Barlier; 41, Joe Skipper; 42, Bill Ingalls/NASA; 44–45, SpaceX; 46, McClatchy-Tribune; 50, Mark Garlick/ Photo Science Library; 51M, AC NewsPhoto; 52, Stephen Chung; 55, Ivan Alvarado; 62, dotted zebra; 63, 3000ad; 66, NG Images; 68–69, NASA/Dembinsky Photo Associates; 71B, Aaron Alien; 73, NASA; 82, Joe Skipper; 88, Ironwool; 91, Veronica Cardenas; 92, Veronica Cardenas; 93T, Veronica Cardenas; 93B, UPI; 96, McClatchy-Tribune; 97T, McClatchy-Tribune; 98T, Shamil Zhumatov; 98B, Geopix; 103, UPI; 105, WireStock; 107, Bob Daemmrich; 108, Gene Blevins; 109, Gene Blevins; 110, Geopix; 112, Gene Blevins; 113, Gene Blevins; 114, Gene Blevins; 115T, Bill Ingalls/NASA; 116–117, Geopix; 118, UPI; 119, Geopix; 120, ABACAPRESS; 121, The NASA Library; 122, UPI; 123, UPI; 124, Geopix; 125, ZUMA Press; 130–131, dotted zebra; 134, Simon Serdar; 137, Blue Origin; 138, Sipa USA; 140, Xinhua; 142, Imaginechina; 144, Xinhua; 147, Xinhua; 148–149, Adrian Mann/ Stocktrek; 152, Xinhua; 154T, UPI; 154B, ZUMA Press; 155, Stockbym; 160, Xinhua; 161, Jason Lee; 162, Xinhua; 164, Xinhua; 165, Xinhua; 168, Ben Von Klemperer; 171, 3000ad.

AP Images: 11; 23; 90, NurPhoto Agency; 94, NurPhoto Agency; 101, NASA; 143B; 153, Ng Han Guan; 163B, Manish Swarup.

Blue Origin: 86–87.

Creative Commons: 31, NASA/MSFC; 71T, TEDxMoscow/CC by 2.0; 77, NASA Glenn Research Center/CC by 2.0; 84, NASA HQ Photo/CC by-NC-ND 2.0.

Getty Images: 10, Cristina Monaro/Corbis Historical; 12T, Cyrus McCrimmon/Denver Post; 30, Joe Raedle; 32, NASA; 35, Bruce Weaver/AFP; 40B, Bruce Weaver/AFP; 99, handout; 143T, AFP; 151, AFP; 163T, Mark Ralston/AFP; 166, Mark Garllick/ Science Photo Library.

NASA: 48, 53T; 60; 61; 65; 67; 72; 75; 78; 83; 85; 100; 102; 106; 115B; 139; 146; 156–157; 157; 158; 159.

NSA Archive: 64.

Shutterstock: 13, James Dalrymple; 14, Mike Mareen; 51T, Vasin Lee; 51B, Jenson; 53B, Simeonn; 54, Sergii Chernov; 57, IndustryAndTravel; 58–59, muratart; 70, Herbert Heinsche; 80, Cristian Cestaro; 97B, EWY Media; 133, Dotted Yeti.

SpaceX: 43.

Voyager Station: 126; 127; 128; 132.

INDEX

First Published in 2022 by Motorbooks, an imprint of The Quarto Group,
100 Cummings Center, Suite 265-D, Beverly, MA 01915, USA.
T (978) 282-9590 F (978) 283-2742 Quarto.com

26 25 24 23 22   1 2 3 4 5

ISBN:978-0-7603-7554-9

Digital edition published in 2022
eISBN: 978-0-7603-7555-6

Library of Congress Cataloging-in-Publication Data available

Design and Cover Image: Landers Miller Design

Printed in Singapore